Make:

TIPS AND TALES FROM THE WORKSHOP

VOLUME 2

A HANDY REFERENCE FOR MAKERS

Gareth Branwyn

Make:

TIPS AND TALES FROM THE WORKSHOP, VOLUME 2

by Gareth Branwyn

Printed in Canada.

Published by Make Community, LLC, 150 Todd Road, Suite 100, Santa Rosa, CA 95407

Make: books may be purchased for educational, business, or sales promotional use. Online editions are also available for most titles. For more information, contact our corporate/institutional sales department: 800-998-9938

Publisher: Dale Dougherty
Editors: Patrick DiJusto and Michelle Lowman
Copy Editor: Sophia Smith
Creative Director: Juliann Brown
Graphic Designer: Terisa Davis
Illustrations: Richard Sheppard

February 2022: First Edition
Revision History for the First Edition
1/14/2022

See www.oreilly.com/catalog/errata.csp?isbn=9781680456370 for release details.

978-1-68045-637-0

How to Contact Us:

Please address comments and questions concerning this book to the publisher:

Make: Community, 150 Todd Road, Suite 100, Santa Rosa, CA 95407

You can also send comments and questions to us by email at books@make.co.

Make: Community is a growing, global association of makers who are shaping the future of education and democratizing innovation. Through *Make:* magazine, 200+ annual Maker Faires, *Make:* books, and more, we share the know-how of makers and promote the practice of making in schools, libraries, and homes.

To learn more about *Make:* visit us at make.co.

To all of the amazing, talented, and just plain decent people I've met in my travels through the maker movement. You give me hope for humanity.

And to Patrick DiJusto — a long-time fellow traveller in tech publishing and the editor of this book — who we tragically lost during its making.

PEOPLE ARE SAYING

(REVIEWS OF TIPS AND TALES FROM THE WORKSHOP)

"Gareth Branwyn is the Tips Master. He scours the workshops of the world for practical, time-saving, life-altering tips to help you **make stuff better, faster, and cheaper.** This book rounds up the best ones he knows."
–KEVIN KELLY, creator of *Cool Tools* and Wired Senior Maverick

"'Tips' is an understated word to describe what's going on here. This book is nothing less than a capstone course in outside-the-box thinking. Even if you never directly use even one of the tips it contains, **you'll come out the other side a much better creative problem solver.**"
–SEAN MICHAEL RAGAN, author of *The Total Inventors Manual*

"I've been meaning to post a review of *Tips and Tales from the Workshop* by Gareth Branwyn, but every time I start, I get distracted by the book itself. I keep flipping through and **learning new things** or being reminded of tricks I once knew."
–LENORE EDMAN, Evil Mad Scientist Laboratories

"It must be hard to write a book like this with such **uncommon clarity** and in so entertaining a fashion as Gareth Branwyn has done here. Gareth clearly has **a deep understanding of making and those who make** because he is a maker himself. *Tips and Tales from the Workshop* is jam packed with invaluable information; it is both a fun read and a reliable shop reference for any do-it-yourselfer."
–ANDY BIRKEY, YouTube maker

"There are **tons of great ideas** here for doing any project better, faster, and smarter. **I'm learning a lot.**"
–BOB KNETZGER, legendary toy designer

A good shop tip is a meme in the original sense of the word (an idea so useful that it quickly spreads from one nervous system to another). Gareth Branwyn has been collecting shop tips for many years, and he has assembled them into a new book called *Tips and Tales From the Workshop*, which is **filled with hundreds of truly useful tips organized by topic**. You'll learn about smart ways to keep track of small parts, plan projects, glue things, mark things, cut things, drill things, and paint things. The tips on 3D printing have **greatly reduced my frustration level**. Even if you don't have a workshop many of the tips here can save you time."
–**MARK FRAUENFELDER**, founder of *Boing Boing* and co-editor
   of *Cool Tools*

"Filled with advice of doing all kinds of stuff. I sat down and started reading it and I lost track of time. **If you're a maker, a crafter, this book should be on your shelf.** It's incredible!"
–**JAMES FLOYD KELLY**, technical book author and YouTube maker

"**This is good stuff**. Trust me!"
–**DUG NORTH**, kinetic sculptor

"*Tips and Tales from the Workshop* is sure to inspire anyone to **get making with newfound ease** and satisfaction. This book embodies the spirit of great mentors, across every medium, and imparts a wizard-like cleverness to its readers. I thought I was clever, and **this book has already prevented at least a dozen new mistakes in my studio**. It's 'ah-ha' moment overload!"
–**BECKY STERN**, Instructables and YouTube content creator

"Whether you're a diehard DIYer or an occasional tinkerer, **this book will make it easier**. I have used his hints to concoct a homemade substitute for a commercial product, drill more accurate holes, make glue work better, and get more utility out of common tools. **I find myself referring to it for advice before starting anything hands-on**. Usually there's a better or faster way than I thought of."
–**ROSS HERSHBERGER**, DIY audiophile and field service engineer

"This is the kind of book you leave lying around, hoping that the plethora of tips slowly drips into your brain for future use. Nothing here is super advanced, so if you're seeking to become an expert anywhere, this isn't for you, but **even the most advanced and experienced shop master will surely find something** here they haven't seen before."
–**CALEB KRAFT**, DIY guru, *Make*: Senior Editor

"Gareth has essentially created a **magic book for makers**."
–**DONALD BELL**, creator of *Maker Update*

CONTENTS

05

CUTTING 43

06

CLAMPING 53

07

GLUING 59

ABOUT THE AUTHOR

Gareth Branwyn is a well-known writer and editor, and a pioneer of both online culture and the maker movement. He is the former editorial director of *Make:*, was a contributing editor to Wired for 12 years, and a senior editor of *Boing Boing* (in print). He has also contributed to *HackSpace*, *Esquire*, *Details*, *I.D.*, the *Baltimore Sun*, and numerous other magazines and dailies. In 1993, Gareth collaborated with Billy Idol on the spoken word lyrics to the opening track of Idol's album, *Cyberpunk*. Gareth is the author and editor of over a dozen books, including the *Mosaic Quick Tour: Accessing & Navigating the Internet's World Wide Web* (the first book about the World Wide Web), *The Happy Mutant Handbook* (with the editors of *Boing Boing*), and *Borg Like Me & Other Tales of Art, Eros, and Embedded Systems*, a best-of collection and "lazy person's memoir," spanning his over 30-year writing career. Gareth is currently a regular contributor to *Boing Boing* and the Adafruit blog. In partnership with Kevin Kelly's Cool Tools network, he also publishes the weekly newsletter *Gareth's Tips, Tools, and Shop Tales*. He lives in Benicia, CA, with his wife, artist Angela White, and their beloved ragdoll cat, Toci.

ACKNOWLEDGEMENTS

It takes a village (to build a tips book). Thanks to everyone in the maker community who contributed to this book, either directly by sending me tips and tools recommendations, or indirectly through your online project documentations and videos where I spotted a technique or tip to include here.

Special thanks go out to the Maker Media team, especially my editors, the late Patrick DiJusto, Roger Stewart (who shepherded this book after the sudden passing of Patrick), and to new books editor Michelle Lowman for getting it across the finish line. To *Make:* creative director Juliann Brown, who is always a joy to work with, to book designer Terisa Davis, copy editor extraordinaire Sophia Smith, and illustrator Richard Sheppard. Also, high-fives to the team at Cool Tools: Kevin Kelly, Mark Frauenfelder, and Claudia Dawson, for keeping the tip fires burning.

Big shout-outs also go to Donald Bell, Jimmy DiResta, Sean Michael Ragan, Kent Barnes, the late Sharon of Figments Made, Andrew Lewis, Andy Birkey, Alberto Gaitán, Dave Mordini, John Graziano, my parents, George and Mary Frances Maloof, and my son Blake Maloof for being such kind and helpful supporters of my work.

Most importantly, I'd like to thank my lovely and talented wife, Angela White, for all of the love, support, counsel, and life-art collaboration.

TIPS CREDITS

As with the first *Tips and Tales from the Workshop* collection, this book was a communal effort. It is the result of seeking out and engaging with makers of all stripes in search of useful ideas to help us all work better and smarter.

The following makers are the sources of the tips found herein. Some of the tips are from videos, online how-tos, and tool references. Others were sent directly from the tipsters themselves. All of these people are amazingly talented makers and almost all of them have websites and YouTube channels. Do a search. Follow them. Having these people on your radar will yield an ongoing and inspired feed of great shop tips, tool recommendations, and project ideas.

AA = Adam Andrukiewicz

AAT = Anne of All Trades

AB = Andy Birkey

ADD = Amie DD

AG = Alex Glow

AGA = Alex Gabriel Ainouz

AkBKukU

AL = Andrew Lewis

AL2 = Aidan Leitch

AS = Andreas Spiess

AS2 = Andreas Salzman

AS3 = Adam Savage

AW = Al Williams

AW2 = Angela White

BC = Bob Clagett

BC2 = Bob Commack

BD = Bill Doran

BD2 = Bobby Duke

BH = Ben Heck

BK = Bob Knetzger

BM = Brett McAfee

BO = Beau Ouimette

BS = Becky Stern

CL = Caroline Lewis

CL2 = Christos Liacouras

CN = Chris Notap

CP = Charles Platt

DB = Donald Bell

DFJ = Dirt Farmer Jay

DH = Dan Hienzsch

DM = Daniel McQuin

DM2 = Dave Murray

DMS = DM Scotty

DN = Dennis Nestor

DT = Derek Thompson

EA = Elisha Albretsen

EK = Emory Kimbrough

EK2 = Eric Kaplan

EM = Ellen Meijer

FB = Fran Blanche

FI – Frank Ippolito

FT = Federico Tobon

GB = Gimme Builds

GC = Geoffrey Croker

GG = Gianfranco
GavazziFisher

GP = Guy Perchard

Habu

Haku3D

HS = Hep Svadja

JB = Jordan Bunker

JB2 = Jack Bonawitz

JD = Jimmy DiResta

JEP = John Edgar Park

JF = Jeremy Fielding

JFK = James Floyd Kelly

JG = John Graziano

JH = Jake Hildebrandt

JM = Jerry Morrison

JM2 = Joe Mayerik

JO = Jay Olson

JP = Jeremy Pillipow

JS = Justin Sparks

JW = Jonathan Whitaker

KB = Kent Barnes

KH = Keith Harker

KK = Kevin Kelly

KK2 = Keith Kelly

KM = Keith Monaghan

KS = Kayte Sabicer

LB = Leah Bolden

LE = Lenore Edman

LF = Limor Fried

LK = Laura Kampf

IS = Izzy Swan

LDO – Linn Darbin Orvar

MB = Mohit Bhoite

MC = Michael Colombo

MC2 = Mike Clarke

ME = Matt Estlea

MP = Matthew Perks

MR = Martin Rothfield

MV = Miguel Valenzuela

NP = Neil Paskin

OS = Owen Smithyman

PC = Paige Cambern

PS = Philip Stephens

QD = Quinn Dunki

RB = Rex Burkheimer

RH = Ross Hershberger

RH2 = Rebecca Husemann

RK = Rahmi Kocaman

RK2 = Roman Khramov

RO = Rich Olson

RO2 = Ryan Oates

RS = Reid Schlegel

SCT = Sara Conner Tanguay

SFM = Sharon Figments Made

SH = Sherry Huss

SH2 = Stefan Hermann

SMR = Sean Michael Ragan

SN = Stumpy Nubs
(James Hamilton)

SR = Steven Roberts

SS – Star Simpson

SS2 = Steven Stoh

SS3 = Sean Sutter

SW = Sophy Wong

SW2 = Scott Wadsworth

TH = Tom Haney

TOT = This Old Tony

TP = Tom Plum

TS = Tim Sway

TS2 = Thomas Sanladerer

TW = Tyler Winegarner

WM = Winston Moy

WO = Windell Oskay

INTRODUCTION
CONTINUING THE CONVERSATION

When I wrote *Tips and Tales from the Workshop*, I was inspired by the story-driven nature of tips and tool recommendations. I wanted the book to not only be a collection of great time-saving tips, clever work techniques, and useful instructions on how to function better in a workshop, I also wanted to tell some of the stories that so often come attached to shared tips and tool recommendations.

What I was delighted to discover was that the book itself became a tool-tale to tell. People liked my book enough to give it to friends and family members. It became an Amazon bestseller. Readers took pictures of themselves holding it and posted them to social media, something I never anticipated. Readers also sent me feedback on the tips they'd applied and stories of how these ideas had impacted their work-lives. And they began sharing new tips and tool recommendations with me.

In response to all of this, I began a weekly newsletter to share new tips I was collecting. Much of the content for this second volume of tips is taken from the best of my weekly tips column on *Make*: (that I ran until 2019) and the weekly newsletter that I've been publishing since.

Like its predecessor, Volume 2 also draws from the best shortcuts, workarounds, and workshop-practices found in the pages of *Make*:, on the Make: Community website (makezine.com), and amongst the wider online maker community.

My great hope is that this second tips collection will further the conversation even more. Please share with me your favorite tips, tool recommendations, and tales from your workspace. I love to hear stories about how you learned a useful technique, how you came by a beloved tool, or about how a project went epically great or horribly wrong. Bonus points for project redemption stories. Let's keep the conversation going!

Gareth
garethbranwyn@mac.com
garstipsandtools.com

HEY, THAT'S MY TIP!

You can't read the comments for many tips videos before someone cries out: "Hey, that's my tip. You stole that from me!" (or something similar).

Tips, like slang, jokes, and funny memes, yearn to be shared; tips want to be free. They are shared from maker to maker, they are seen in online project documentation and in videos and are added to the arsenal of shop techniques by those exposed to them. They end up in tips articles and books. Tips are promiscuous. And that's a good thing.

I've tried, as much as possible, to at least retain the sources of the tips collected in this book. This attribution is not necessarily the author of the tip, but just the person I got it from. These credits are displayed as initials at the end of tip entries and there is a list of these contributors in the front of the book. In all cases, the descriptions of the tip (unless quoted) are mine. If I missed any tipsters, I apologize.

RUST NEVER SLEEPS

In the original *Tips and Tales from the Workshop*, we used the design element of a benchtop to tell the visual story of a busy workshop and the projects that have moved across it. Like workbenches, tips and tools, passed from one person to another, also carry the marks of the people that have used them.

Another visual story that can be "read" upon the made world is the patina, wear, and rust that time and use accumulates on the tools that we use and the objects we build with them. Rust is time's artform. As rusting happens, signaling the impending demise of an object, it does so while creating something of beauty in the process.

In the 1990s, sci-fi author (and former *Make:* columnist) Bruce Sterling created a design "movement" he dubbed Viridian Green. One of the main tenets of Viridian Green was: "Embrace decay." (The movement also embraced sustainable technologies, smart design, and global social and environmental awareness.)

In a world obsessed with youth, what's next, and the latest shiny object, there's something noble, corrective, and smart in embracing the old and the perennial, and breathing new life into the cast off and the exhausted. To celebrate such ideas, for this second volume of *Tips and Tales*, we chose the design element of rust to beautify the pages of this book.

MAKING ASSUMPTIONS

In deciding whom I was talking to in this book, I assumed that the reader is already an avid maker of some stripe and has a basic working knowledge of the tech, terms, and techniques in these tips categories. Acronyms are always unpacked the first time they're used, and, when appropriate, tech terms are parenthetically defined. If you find yourself encountering a word or concept that you don't understand, the internet is your on-demand learning machine.

For you, the reader, don't assume that because something is in this book that it has been thoroughly tested by me. Try these ideas out on your own, with a healthy dose of skepticism and always with your own safety in mind.

Some tips will not work as satisfactorily as you'd like, some will work just fine but won't fit into your workflow and so they won't stick for you, and others will be real game-changers and you'll wonder how you ever got along without them. This is all the nature of the beast.

In exploring these and other tips online, you'll find that there is often loud naysaying from trade professionals who will tell you never to use a tool, material, or process. Don't let that stop you from experimenting. As long as you're working safely, try something and see how it works out for you. Do your own research and see what others online think about the efficacy of the idea. Shop tips and new techniques are the laboratory of making. Be a good empirical scientist and test things out for yourself.

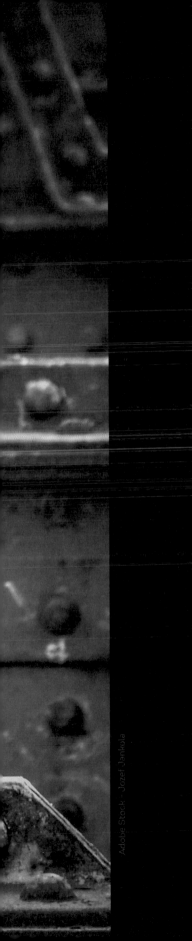

1
PLANNING AND META TIPS

I have always been fascinated by the radically different ways that humans plan. Through time and experience, each one of us develops a (hopefully) workable way of getting things done. Every person is a planner, and they have a planning method, whether they acknowledge it or not. In my own work and life, I've tried countless planning and to-do systems over the years. None of them have stuck for long. Over time, I've just accepted the fact that I'm a haphazard planner and that's just how it is. I pick and choose from various planning "systems" and my approach changes over time. One thing I've consistently found is that it's helpful for me to expose myself to new ideas around planning and new personal information management systems. That's why I love collecting tips related to planning. Here are some recent ones that I've been inspired to keep around.

In this volume of *Tips and Tales*, I've also added what I call "meta" tips. Sometimes, the tips that have the biggest and most lasting impact on your work and personal life are not particular to a technique or procedure, they are more general and relate to how you think about your work and your approach to it. Chapter 1 starts off with some of these meta tips.

⭐ KNOWING WHEN YOU'RE ON TO SOMETHING

I took a weekend seminar once on bringing your ideas to market. The teacher shared something that I've never forgotten. He said, when you think you've come up with a truly new and innovative idea, and you share it with others, pay attention to how quickly and easily they embrace it. If they immediately get excited, you may be on to something, but likely, others are already on it, too. Immediate and enthusiastic reactions mean that the cultural and market vectors may have already converged on the very thing you've just identified. If, however, you tell others and they don't get it right away — but after thinking about it, or you do more explaining they get it then (i.e., a slow uptake rather than an immediate eureka!) — you just may be ahead of the curve. When I first heard this, some 30 years ago, the tip itself worked in exactly the way that it describes. It didn't have tremendous resonance to me in the moment. But, over the years, it has been borne out time and time again.

⭐ IDEA FISHING WITH DAVID LYNCH

Filmmaker and multidisciplinary artist David Lynch says: "Ideas are like fish. If you get an idea that's thrilling to you, focus your attention on it and these other idea-fish will swim to it. It's like bait. They'll hook on to it and you'll get more ideas. And you just reel them all in."

⭐ DETAILS LAYER

Scott Wadsworth of the YouTube channel *The Essential Craftsman* has a wonderful statement that speaks to something very important in all aspects of making (and in life): "details layer." In every step of a project or activity, the precision with which you do one step, one layer, carries over into the next, and the next, and the next. Over time, mistakes and imperfections compound, so it's important to do each step as well and as thoughtfully as possible. **[SW2]**

THE UNIVERSE IS A COLLECTION OF PARTS

I met an inventor once named Perry Kaye. He had a brilliant approach to prototyping his designs. He didn't try to reinvent the wheel – he used existing wheels from something else! He called this approach "Frankenstein prototyping." When Perry came up with a new idea, rather than going the conventional route of drawing up plans and paying a rapid prototyping service or someone else to fabricate it, he'd just head to Home Depot, Toys "R" Us, and the local hardware store. He'd find the parts he needed on existing products (a handle here, a type of blade there, this motor, that gearbox). Then, he'd cut up these existing products and stitch them together into his new monster creation.

This is an incredibly powerful perceptual shift — to see the physical world around you as a collection of parts that are currently in one configuration, but are just waiting to be taken apart and recombined into something new. Especially with today's 3D printers and cutters, high-performance adhesives and other materials, and so many cheap components readily available online.

Besides saving time and money, there's an added benefit. When you've spent so much time, money, and effort prototyping an idea, you become literally invested in making it work, even if it doesn't. But when you've only invested an afternoon and a few bucks on a Frankenstein prototype, you're more likely to salvage whichever parts you can, and move on to the next idea. This method of rendering your ideas allows you to iterate quickly and gets you to a smarter, more viable design much faster.

Of course, you don't need to be an inventor in the classic sense to benefit from this way of looking at the world. You can make one-off creations with this method, or solve vexing design problems on existing projects. We have this perceptual blindness where we tend to see things as they are rather than the potential for what else they can become. Frankenstein prototyping is a way of training oneself to see that potential.

★ ART AND ENGINEERING NEED EACH OTHER

Steven Roberts, the "high-tech nomad," dropped a quote many years ago that has become a central pillar of my approach to thinking and creating: "Art without engineering is dreaming. Engineering without art is calculating." Similarly, the cyberneticist Gregory Bateson talked about the need for a balance of "rigor and imagination" in one's endeavors. Too often, I think there's a tendency to overdevelop one of these muscles and not the other. I started out on more of the dreaming side of things and have spent much of the last three decades learning to balance that with a deeper, hands-on understanding of the physical sciences side of things. Where do you fall in this mix? What part of this equation do you need to work on more? [SR]

★ MAKING IT PERFECT ENOUGH

Andy Birkey is a maker who does amazing Gothic church restoration projects. Andy says that when you start a project you want to try and do everything as perfectly as possible, but as the deadline draws closer, you need to start making an important pivot in your mind — letting go of the quest for absolute perfection. It's time to start focusing on making it "perfect enough." As an example, he points out that he's been up in the high ceilings of old churches. Up there, you can see all of the imperfections that previous builders left because they knew they'd never be seen from the ground. The ceiling looks amazing at the distance at which it is meant to be viewed. It doesn't need to be perfect. It just needs to be perfect enough. [AB]

★ THINKING OUT LOUD IN A CLOUD DOC AND THEN INVITING FRIENDS INTO IT

I love all of the diverse ways people use online collaboration and communication tools. On Twitter, maker extraordinaire Star Simpson shared how she uses cloud-based docs to post thoughts and then she invites people into the document whose feedback she's after. "I've been doing this kind of 'lazy blogging' [for a while] where I just write a bunch of thoughts in a Dropbox Paper doc and then text it to the friends whose input I want on it. I can see who's accessing, deleting or editing, and there's no dealing with a content management system or trying to structure/order ideas." **[SS]**

⭐ HUG REPORTS

"Hug Reports are the opposite of Bug Reports," says Limor Fried of Adafruit Industries." A Hug Report gives everyone a chance to say thank you to someone during a group meeting. Little things matter, and if we're all celebrating each other, it makes your work and your values better. And that's something our community and our customers notice in how we do things." For the Hug Reports, employees single out someone who's done something special, admirable, above and beyond the call of duty. This can be as general as "I like how so-and-so is running their department" or about a specific action that impressed a member of the team. **[LF]**

⭐ DELAYING YOUR THANKS

This one's a little on the sneaky side. When you ask someone to do you a favor online, and they say they will, don't respond with your "thank you" right away. Wait to see if they do what was requested. If they don't (within a desired time-frame), then send your "thank you." It will serve as their reminder. This way, you won't risk annoying them by sending additional nagging emails.

⭐ DIVIDING UP INTIMIDATING PROJECTS

If you have an intimidating project, like a complicated piece of flat-packed furniture that you are dreading putting together, don't try and tackle it all at once. Do the assembly over multiple days and just work on it an hour (or so) at a time. Dividing it like this will keep your spirits up, keep you relaxed, and the results will probably come out better. This, of course, is only possible if you have the space to do this, your partner or housemates are okay with it, etc.

MAKE IT SEEM HARD, NOT BE HARD

Years ago, I was at a Bay Area Maker Faire. I'd been teaching under our new "Learn to Solder" tent. Participants paid a couple of bucks, sat down, learned the basics of soldering, and then put together a small electronics kit. The kit, a flashing-LED Maker Faire pin, was little more than a tiny printed circuit board (PCB), a blinky LED, battery holder, battery, and pin. The kit had about five solder points. I soon came to realize why the workshop and that kit were such genius. They made an intimidating skill seem simple and accessible, and the kit was just challenging enough to feel like you'd accomplished something. It was a literal badge of accomplishment you could wear. There was a very important idea here: Make it seem hard, but not be hard.

This was immediately driven home after the workshop. I was at the Solarbotics booth, talking to Dave Hrynkiw. A father and his maybe 10-year-old daughter were looking at kits, "Learn to Solder" badges blinking away on their shirts. Dave began showing them kits, emphasizing solderless beginner kits. The girl, slightly indignant, said: "NO, I don't want solderless. I know how to solder," pointing proudly at her badge. They bought a solder-based kit and off they went.

When teaching something new, create the shortest distance between the learning part and the completion of a successful first project. If the student can wear it proudly, or otherwise show it off, all the better.

⭐ PUTTING TOGETHER AN INSPIRATION BOOK

While working on the 1992 film *Bram Stoker's Dracula*, Francis and Roman Coppola put together a sort of inspiration bible (I call mine "muse books") of art and photographs that represented the mood they were trying to invoke in the film. Whenever they got creatively stuck or needed some inspiration, they'd poke their head in their muse book and take a big whiff. I start many of my projects, from book writing to painting gaming miniatures, by putting together a collection of images and ideas that fire up my imagination.

⭐ A MATERIALS SAMPLES BOOK FOR YOUR SHOP

If you're going to take the time to acquire materials samples and swatch books, you want to make sure you have them organized and accessible for future use. The guys at Tested.com keep materials sample books in binders, ready for reference. Many suppliers are more than happy to give away free (or for a modest fee) samples of their products. Don't be afraid to ask. You can learn more from this Tested video: youtu.be/YO9ET7bpE1A.

⭐ ADVICE FOR YOUNG MAKERS

Sherry Huss, the "mother of the maker movement," put together this little "manifesto" of top-level tips for instilling the spirit of lifelong learning, adventure, and making among young people. You can download the PDF here: bit.ly/2Ezg9uB. **[SH]**

A MANIFESTO FOR YOUNG MAKERS

1	**MIX IT UP** — Work **outside** of your area of interest and **comfort zone**. Experience as much as you can with a **diversity** of people and groups.
2	**DO THE MATH** — Learn to work with numbers as soon as possible – whether you are **making** something or managing money. Extra points for learning the metric system.
3	**PUT YOURSELF OUT THERE** — **Attend** meetups. **Speak** in public. Join the debate or drama club. Share your ideas. Find your voice. Build your global network. **Engage.**
4	**TRAVEL UNKNOWN PATHS** — Develop a **new skill.** Learn a foreign language. Play an instrument. Grow or cook your food. Learning is a lifelong adventure.
5	**BE PRESENT AND ACTIVE** — Get and stay **involved** in community groups – be it community service or special interest groups. You will pick up new skills, collaborators, and mentors along the way.
6	**GET HANDS ON** — **Work with** as many physical **tools** as you can. This is neccessary for learning and making. Include technical, design, and fabrication tools in this process.

Sophy Wong

⭐ THUMBNAIL AND STORYBOARD TEMPLATES

Designer, maker, and author Sophy Wong laser-cut these templates on her Glowforge so that she could quickly create thumbnail and storyboard pages in her design notebook. You can download the files to laser-cut your own at sophywong.com. **[SW]**

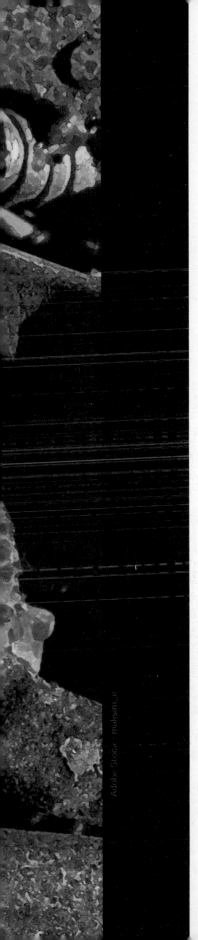

Adobe Stock – matsim_e

2
GENERAL ORGANIZING

For many of us (most of us?), organization is often an aspirational thing. We read books on the subject, watch YouTube videos, read articles, longing to live the tidy, seemingly happy and more content lives enjoyed by "organizing consultants."

But let's be real. Life is messy. As the saying goes, you can't make an omelette without breaking some eggs. The process of living is a dynamic balance of order and chaos, order and disorder. And nowhere is this more true than when you're neck-deep in a project. The idea is not to always stay squeaky clean, but rather, to take enough time to create some decent overall organization in your life so you can give yourself permission to get dirty — to make a great and terrible mess — when inspiration (or a deadline) strikes.

Hopefully the following basic organization ideas and tips will help you better ply the rising and falling tides of (dis)order in your life and in your workflow.

⭐ BEING KIND TO FUTURE YOU

Uber-maker Adam Savage has an adage that he often repeats about being kind to your future self. This is about keeping yourself organized and about cleaning up after yourself. Going that extra organizational mile, and making sure everything is put back where it belongs after doing a task, he says, is giving a great gift to future you when you return to do the next project and everything is where it should be. **[AS]**

⭐ CLEAN UP AS YOU GO

Adam Savage also recommends cleaning up a complex, messy task as you go. This serves several purposes. You are not only cleaning away the clutter, making it easier to stay focused on the task at hand, but switching gears and cleaning for a bit is a great way to clear your mind and come back to the task after a few moments of distance from it. **[AS]**

⭐ CHRISTMAS LIGHTS STORAGE TIP

You can make productive use of your empty 3D printing filament spools by using them to spool up and store your holiday lights (or anything else that can benefit from spool storage, for that matter).

"DREAM IT, MAKE IT" WITH SHARON OF FIGMENTS MADE

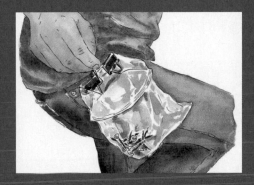

In 2018, I was doing a regular column on *Make:* called *Weekend Watch*, where I'd profile a largely "undiscovered" YouTube maker in an attempt at giving them wider exposure. One of the people I profiled was a woman from New Jersey who simply went by the name of Sharon, or more often, Sharon of Figments Made (FigmentsMade.com).

In doing research for the piece, I came across a tip on Instagram that she'd shared: "That time you didn't bring your tool belt and needed to have a bunch of drywall screws handy." The post included a photo of Sharon with a plastic sandwich bag, filled with screws, binder-clipped to her jeans. I posted the photo and tip to my weekly tips column. Sharon was thrilled by both the *Weekend Watch* piece and the tips entry. We became online friends after that, with her often sending me tip ideas and links to maker channels she thought that I should have on my radar.

On September 29, 2020, Sharon tragically died in her sleep. It was a huge shock to the maker world. In the profile that I'd done, Sharon had talked about two things — how much she loved making people laugh with her project videos and how much she loved being part of the maker community. And what an active part she was. She was always promoting other people's content, sharing ideas, and giving encouraging feedback to project posts on Instagram. She was a loving, sweet, and supportive presence within our community.

The motto of Sharon's channel was, "Dream It, Make It." She obviously tried to live by that edict, with her expansive creativity and numerous whimsical creations. May we all try to do the same.

I had the wonderful illustrator for this book, Richard Sheppard, do a watercolor version of Sharon's hip pocket baggie "hack" (seen above). In honor of her memory, it also graces the cover of this book.

★ USING STYRO TO ORGANIZE HARDWARE DURING ASSEMBLY

If you're disassembling something that has a lot of screws of different types and sizes, consider repurposing a piece of scrap styrofoam to temporarily house and organize the hardware as you remove it. Just press each piece into the styro. You can even

★ WHEN TO BUY AND WHERE TO STORE SPECIALTY TOOLS

On a Q&A video on Tested.com, Adam Savage was asked: "What is your favorite tool that you rarely use?" His answer: Alligator forceps. His general rule is that, if he has a need for a specialty tool like this more than three times in a year, he buys one. In terms of storing low-frequency-use tools, he has a great approach. He asks himself: "If I didn't have it right now, where would I look for it?" And that's where he stashes it. He tries to not get clever, to not overthink it, but rather, he goes with the first place that pops into his head. **[AS]**

★ GENERAL SEARCHING TIPS

This one is from Kevin Kelly of Cool Tools: When you want to learn about something, do an internet search on the desired subject and then add "forum" (or "solution"). This will likely scoop up all of the online communities where that subject is being discussed, worked on, and troubleshot by a passionate group of explorers. **[KK]**

⭐ TOSS AWAY YOUR USER AND SERVICE MANUALS

We all have those drawers, in our kitchens and workshops, filled with user's manuals, service manuals, spare parts order forms, warranty cards, and other paperwork that came with our tools and consumer electronic devices. We almost never refer to them, and they end up clogging up drawers that could be put to better use. Most user and service manuals, parts lists, and the like are available online now. I can't remember the last time I looked up such documentation and didn't find it, even for antique machines and vintage computers. Do yourself a favor, look up the manuals for the devices in those dreaded drawers and throw away anything that's available online.

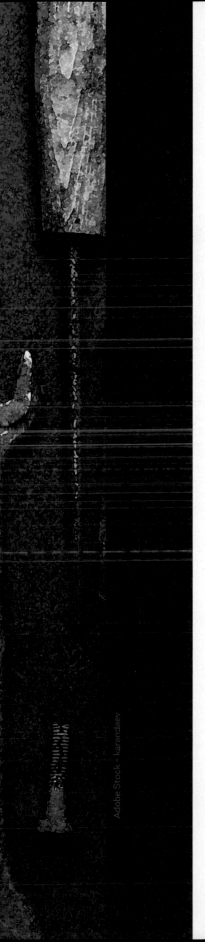

SHOP ORGANIZATION

A clean shop is a happy shop. An organized space means an organized mind. Messy people are more intelligent. Messy people are more creative. Etc. Etc. We've all heard the clichés, often at odds with one another about whether organization or chaos is better for creative self-expression. Obviously, it comes down to personal preference.

If you're the kind of person who can't focus, who can't stand clutter while working, keep your space clean and organized. If you're the type whose workflow is seriously impeded by stopping to clean up as you go, then don't. Everyone's style is different. The important thing is to identify what you're comfortable with and try to serve that style.

All that said, there are lots of things you can do to help optimize the layout and workflow of your workspace regardless of how you roll. The following tips in this chapter will help keep both organized and more... um... exuberant makers happier and more productive in the shop.

⭐ EVERYTHING ON QUALITY, LOCKABLE CASTERS

As tool needs and priorities change, it's nice to be able to reconfigure your setup. Put all tables, storage cabinets, supply carts, etc. on lockable casters. And don't be tempted to get cheap casters. Buy the highest quality ones you can afford.

⭐ STORAGE CASES WITH REMOVABLE BINS

If lots of your components, hardware, and other shop bits and bobs are small, consider investing in plastic portable storage cases with removable bins inside. You can find these at online tool warehouses for under $6-10/each. I bought several dozen during my shop re-org and still have a few in the closet if I need more.

⭐ STORAGE CASE RACKS (ON WHEELS)

It's common for those who use the above storage cases as a central part of their storage tech to build simple wooden carts on casters to house them. Having a number of these with parts and hardware well-organized in the cases is a fantastic way of pulling in the components you need when you need them (and then pushing them out of the way when you're done).

⭐ OVER-LIGHT THE SPACE

You need light. Lots of light. More light than you think. Thankfully, good shop lighting can be very affordable these days. High intensity, linkable LED utility shop lights can be found online for under $130. For desk lighting, I use LED panel lights (created for video recorders) attached to goosenecks.

⭐ NO-TIP SHOP BOTTLES

Want to ensure that the bottles of lubricants, glues, solvents, etc. in your shop can't get easily knocked over? Glue them to a wide, sturdy base. For a truly knock-proof bottle, glue them to a scrap of metal sheet stock. **[Habu]**

⭐ SUBSCRIBE TO GROUPS THAT INSPIRE AND ENCOURAGE YOU

Subscribe to Facebook groups and YouTube channels that encourage shop organization and clever storage solutions. I follow the Shop Hacks group on Facebook and frequently get inspired by what members share there.

⭐ PICK A DAY EACH WEEK FOR A SHOP REBOOT

Organize your shop weekly. Pick a day of the week, say Sunday, and get in the habit of using the end of that day to clean and organize your shop so that you always begin the work-week with a fresh reboot. It's also a good time to take stock of your supplies and materials to see if you need to re-order anything. I find this weekly ritual helps to get things off to a clean and optimistic start.

⭐ SHOP NAIL SALON TECH

It may surprise you, but a great place to find storage solutions for your shop is in the nail salon "aisle" of your favorite online mega-retailer. Nail polish racks, shelving, polish mixing machines, and even nail art paint brushes are all useful in your shop. And most of the products are surprisingly cheap. I spent $40 on a laser-cut storage rack for acrylic hobby paints and it took an hour to assemble it. I got a clear acrylic nail polish rack for half that price, it took about five minutes to put together, and it holds more paints!

⭐ PEGBOARD TOOL SHELF

If you have scraps of pegboard, you can mount them on pegboard brackets as a shelf and even use the peg holes to organize drivers and other similar tools. **[Via Family Handyman]**

SHOP MEMORIES OF MY FATHER

My dad had jars of miscellaneous bolts, screws, etc. He always had a set of old newspapers tucked into the corner of his workbench. When he needed a bolt, screw, or nut, he got the appropriate jar down, unfolded the newspaper, and poured out the jar's contents onto the newspaper. After he found the hardware he needed, he just folded the newspaper up and poured the contents back into the jar. It was very efficient and neat.

My favorite workshop memory from my childhood was the summer we built a small, two-story playhouse in the backyard. My late dad was the manager, but us boys did all the work — sawing, nailing, etc. We did it during that summer's evenings. Dad took advantage of being in the garage to smoke a cigar, his venerable old Zenith tube radio (that he'd had since he was a kid) on the corner of the workbench, usually playing a Detroit Tigers baseball game (we lived about two hours from Detroit, and were within perfect reception range of WJR AM 760, a powerhouse AM broadcaster). Dad got to teach us basic woodworking — we used hand saws mostly, though he did teach us how to use the table saw and drilling (manual and electric). It was a lot of fun, and I can still picture it as I write this.

My favorite current workshop memory is of when we moved into our previous house; my late father-in-law found a stash of solid core doors that had been left behind in the basement. He used one of them to make a workbench, using some old steel workbench legs that he had stashed away. It was sturdy, and serviceable, and even though now, a couple of decades later, I could easily afford a bigger and better bench, I can't bear to part with it. I think of him every time I use it.

I passed that basic familiarity with tools and techniques to my daughter, who isn't afraid to tackle most small projects. She's handier with tools than most of her friends, and especially her new boyfriend, who takes it with good grace and lets her teach him.

—Steve Stroh

⭐ HARBOR FREIGHT TOOL RECOMMENDATIONS

On the YouTube channel *Real Tool Reviews*, Daniel McQuinn does serious, in-depth reviews of tools, as well as product-vs-product shoot outs. In many of his videos, he takes a critical look at Harbor Freight tools, and discusses which of their products are actually great values. Two items that he highly recommends are the U.S. General 5-Drawer Mechanic's Cart (which can be had with a coupon for as little as $190) and the Braun Slim Bar LED Light (as low as $20 with coupon). I have always contended that Harbor Freight gets a bad rap and that you just need to know what you're getting and which tools to steer clear of. Channels like Daniel's can help. **[DM]**

⭐ TIN CAN WIRE ORGANIZER

Make: contributor Andrew Lewis made this handsome little wire organizer from a tin can and some 3D printed parts. Andrew has the 3D design files online. And you can resize the spools and end pieces to fit different size cans. You can download the .STL files at lewis-workshop.com/ downloads/spool-can.zip. **[AL]**

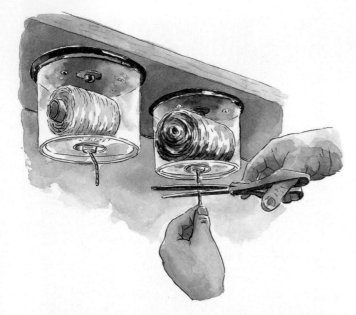

⭐ STRING AND CORD DISPENSER

I love this idea from from *Family Handyman*'s YouTube channel for creating shop string and cord dispensers using old blank CD/DVD containers.

⭐ WORK THAT CUBE!

If you have rafters in your shop, consider using brackets to hold materials in the spaces between the rafters. Or, you can build (or buy) pull-down shelving that folds up into the rafters like attic stairs. You should also think about your wall space (and even the backs of doors) and how you might best utilize them. Go vertical! Work the cube!

Jake Hildebrandt

Andrew Lewis

⭐ ORGANIZING CABLES IN DOLLAR STORE PENCIL CASES

My friend Jake Hildebrandt shared this idea. He uses dollar store zippered pencil cases to organize and label all of his cables in a drawer. **[JH]**

⭐ TURNING ANGLE STOCK INTO STORAGE SHELVES

Andrew Lewis created these 3D printable brackets that can turn custom-cut lengths of 25mm metal angle stock into storage shelves. You can download the .STL files here: lewis-workshop.com/downloads/paintshelf-corner-pair.stl. **[AL]**

⭐ HEATSINK STORAGE RACK

Jake Hildebrandt shared this: "An old heatsink makes a great tweezer storage!" [JH]

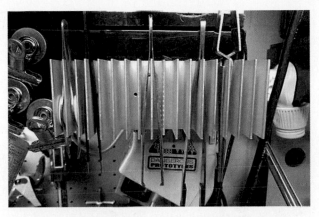

⭐ COVERING YOUR WORK TABLES

One great way of keeping your workbenches cleaner and neater is to always cover them with kraft paper before you start on a new project. You can get rolls of paper extremely cheaply. Use 4" masking tape to seal the paper to the table. Changing the paper after each major job also makes for a nice ritual of cleaning, putting tools away, and getting ready for the next job. [AW]

⭐ NO-ROLL PENCILS

Here's a great WDITOT (why didn't I think of that?): There are a number of marking tool designs and tool modifications that address the problem of pencils, pens, markers, and scribing tools rolling off of workbenches. And then there's a little flag of masking tape. [Via *Family Handyman*]

⭐ MAGNETIC PICK UP TOOL

YouTuber Dirt Farmer Jay has a different approach to picking up loose metal parts than Jimmy DiResta's magnet inside of a bandana trick. Jay has made a set of magnet pick up tools that he keeps handy throughout his shop. These are little more than chunks of scrap wood with magnets glued to them. You can use these to "vacuum" up loose hardware and other metal pieces on your bench or to pick up metal objects that have fallen in hard-to-reach places. [DFJ]

⭐ BLACKBOARDING THE CLOSETS

Spray the cabinet doors in your workshop with chalkboard paint (readily available in arts and crafts stores and online). You can use the chalkboard to keep track of what's in the cupboard, or use it instead of scraps of paper to make notes about your projects, measurements, resistor values, and so on. **[AL]**

⭐ PENCILS, PENCILS EVERYWHERE

Federico Tobon read about keeping pencils everywhere in the shop in the first *Tips and Tales from the Workshop*. So, he made a set of chipboard pencil holders to place at workstations throughout his school's makerspace. They were laser cut and designed using boxes.py (festi.info/boxes.py). So great to see tips being put to use and inspiring people like this. **[FT]**

⭐ KITTY LITTER RUST PROTECTION

To prevent the rusting of your tools, put a scoop of crystal kitty litter into a square of cloth and tie it into a bundle. Place the bundle in your toolbox and the litter will act as a desiccant to remove moisture and help prevent rust. You can also use packs of silica gel that come with many products these days to fight moisture. **[*Family Handyman*]**

⭐ MAKING A SIMPLE, EFFECTIVE ROUTER BIT HOLDER

This is one of the coolest quick-builds I've seen in a while. Keith of *Keith's Test Garage* created three bits organizers for his rotary tools and router bits using dado cuts to create inter-sectional slots to hold tools with ⅛", ¼", and ½" shanks. He got up to 384 ⅛" spaces (for rotary bits) after only minutes of cutting dados on a table saw. Here's a video of the project: youtu.be/WPQ__b27jyM. **[KK2]**

⭐ CHRISTMAS TREE REMOTES IN THE SHOP

Jerry Morrison posted this in the Shop Hacks group on Facebook: "I've been using Christmas light remotes from the big box store for over a year to control my shop vac and my Jet dust collector." You can often get these remotes on deep discount after the holidays. They are a simple on/off remote and wall plug. You can plug tools into them that you want to be able to turn on and off at a distance. **[JM]**

⭐ CREATING A DUCT TAPE CONTAINER

Maker Eric Kaplan writes: "Today I discovered that you can use 3"
PVC pipe 'test caps' to cap off a roll of duct tape. They work really
well for this application. The resulting storage well holds about the
same volume as an Altoids tin. The caps cost around 40 cents
each." **[EK2]**

⭐ SIMPLE CYCLONE DUST COLLECTOR

Chris Notap created an effective cyclonic dust collector (a device that uses centrifugal force to remove dust particulates from the air) using little more than two Home Depot buckets and some plastic piping and fittings. I love the way he created a friction fit with the two buckets mouth-to-mouth by simply beveling the edge of one bucket. Video here: youtu.be/1WnitgYFnE0. **[CN]**

MEASURING AND MARKING

For many makers, measuring and marking can be an intimidating part of a build process. If you make a mistake in measuring, or even worse, a mistake in cutting, you can ruin a project (or at least have to measure, mark, and cut a new workpiece). So, measuring and marking are very important.

But here's a heretical thought. Sometimes, you don't need to measure at all. One of the keys to moving on to a greater degree of relaxation around measuring and marking is to nix the math altogether. Obviously, this is only possible in situations where the final project doesn't need to be precise and the dimensions are up to you. But it's good practice to try eyeballing and freeforming on projects that allow it. Try doing some projects with as little measuring as possible. Think in terms of proportions and relationships rather than numbers. Use stop blocks (a simple jig clamped down on your work surface) for repetitive cuts of the same size. Use one piece as a template to cut other pieces of the same size. Improvise where you can. Again, this is only possible under certain conditions, but along with getting good at measuring and working with precision, working your more improvisational muscles is also a good idea.

Regardless of whether you're looking for precision or better ways to wing it, this chapter should reveal some useful takeaways.

⭐ YOU RULE

In the first volume of *Tips and Tales from the Workshop*, I included a tip about knowing the measurements of different parts of your hand, arms, and feet so that you can use them in a pinch for rough measuring when no ruler is at hand. Here's a worthwhile addition. Dump measured amounts (e.g., teaspoon, tablespoon) of something in the palm of your hand to get a rough idea of what those quantities look like. For anything that only requires rough measuring, this will give you an adequate approximation. In the above picture, I have a tablespoon- (left) and a teaspoon-worth of flour (right) in my hand.

⭐ FINDING THE THICKNESS OF A WIRE

If you need to find the thickness of a wire but don't have a micrometer or calipers handy, wrap the wire around a dowel many times in a tight helix leaving no gaps between the coils. Now, just measure the width of, say, thirty coils with an ordinary ruler and divide by thirty. The more coils you wind, the more accurate your measurement. And, even if you do use top-quality digital calipers, it's even more accurate if you use this wind-and-divide method than if you measure a single thickness. **[EK]**

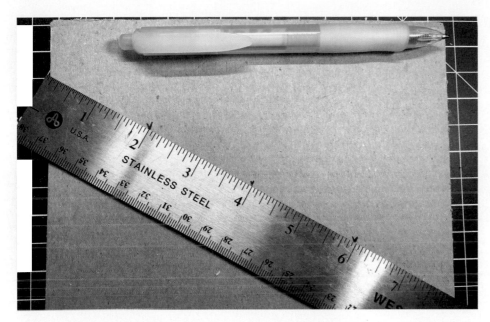

⭐ EASILY DIVIDING A WORKPIECE

To easily divide a workpiece into equal parts, do the following. Let's say you have a 7" wide piece of cardboard that you wish to divide into four equal strips. Place your ruler so that the zero mark is along one edge of the workpiece. Now, move the ruler down along the other edge of the cardboard until you reach a number that is easily divisible by four (in our case, 8"). Next, mark the 2", 4," and 6" ticks on the ruler. Do the same farther down on your workpiece and connect the marks perpendicular to the sides of the cardboard. You now have the piece divided into four equal strips.

⭐ FINDING THE CENTER OF AN OVAL

Here's an easy way to find the center of an oval: Perhaps you have a wooden plaque, metal sign, or a frame that's an oval shape. Place it onto a sheet of paper and trace its outline. Cut out the oval. (Or, place the oval onto the paper and cut around it with a hobby knife.) Fold your paper oval neatly in half across its short axis, then in half again across its long axis. (In other words, carefully make two perpendicular folds to create a quarter oval.) Finally, unfold the paper oval — where the two creases cross is the center. Now you have a template to mark or drill the center of your oval workpiece. **[EK]**

⭐ DIVIDING A CYLINDER INTO EQUAL PARTS

If you want to divide a cylinder into equal parts, first wrap a piece of masking tape around the circumference. Now, remove the tape, place it on a flat surface, and measure and mark your division (four equal parts, five equal parts, six, etc.). Now peel up the tape, reapply it to the cylinder, and transfer the tape marks onto the cylinder. From Laura Kampf's measuring and marking video: youtu.be/MiOWqjewYxw. **[LK]**

⭐ ADDING BLADE "KERF" WHEN MEASURING CUTS

If you're measuring out a series of same-length cuts in a piece of wood, don't forget to include the saw blade kerf (the width of the blade) in your measurements. So, for instance, if you're cutting a series of 10" long pieces for a 2×4 board and marking them all ahead of time, the first mark would be at 10", the second at 10⅛", the next measurement would be 10¼", then 10⅜", and so on.

⭐ "BORROWING" A TAPE FROM THE HOME STORE

If you get to a hardware or home store and realize that you've forgotten your tape measure, don't fret. Borrow one of theirs, do your measuring, and put it back.

⭐ THE RIGHT MARKING TOOL FOR THE JOB

When marking, use the tool appropriate to your materials, tolerances, and working conditions. For example: thin mechanical pencil for finishing work, carpenter's pencil for framing, masonry, etc., where tolerances are low (and for better visibility), and grease pencil or permanent marker for indelible marks, especially outdoors.

⭐ MARKLESS MARKING

If you're making a lot of marks on a workpiece and don't want to have to clean those marks up afterwards, put down masking/painter's tape where the marks are likely to go. When you're finished, remove the tape. All clean! **[EA]**

⭐ STOREY STICKS

If you do a lot of repetitive measuring, it's a good idea to create what's called a storey stick or storey pole. On a narrow length of board, you mark whatever custom measurements you want to remember. The name comes from homebuilding and the use of such a marking stick that is one storey high. Most people use pieces of wood to create storey sticks. But another great idea is to use an old carpenter's folding ruler. That way, you'll have your commonly used marks and you'll have regular measuring capabilities.

⭐ USING A CARPENTER'S PENCIL

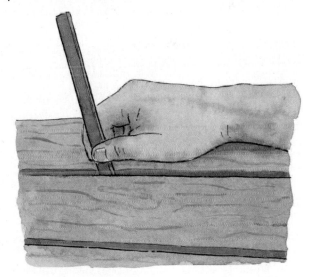

In a video on *See Jane Drill*, Leah Bolden runs through some of the features of a carpenter's pencil that you may not be aware of. Like, did you know that a carpenter's pencil has a set width and thickness (½"×¼") and is designed to be used for things like deck board spacing (at ¼")? You can cut the ends of a carpenter's pencil into different lead shapes for different types of marking? See more here: youtu.be/GAsSOnU0jqk. **[LB]**

⭐ USING PENNIES AS SPACERS

On the Shop Hacks Facebook group, Adam Andrukiewicz from A-Z's Custom Woodworking wrote: "If you ever need to ensure consistent, even spacing of multiple gaps on a project, a US penny is ¹⁄₁₆" thick. Used on this occasion in making a jewelry box to ensure equal and even spacing above, below, between, and on both sides of the jewelry box drawers so that all the reveals are ¹⁄₁₆" all around." BTW: You can also use playing cards as spacers. They are usually made from 11.5 pt card stock, which is .0115" thick. You can use them in multiples to get the spacing that you desire. **[AA]**

⭐ TESTING SQUARES FOR ACCURACY

How to tell if your combination square is true or not: Using a marking knife (or other marking tool), scribe a line from one edge of the board and then flip the square to scribe over the same line from the other edge of the board. You should end up with a single scribe-line of a consistent width. See this video (at 14:50) on the *Stumpy Nubs* YouTube channel for a demo: youtu.be/BhFIZQvywMw. **[SN]**

⭐ RE-TRUING A FRAMING SQUARE

If you have a carpenter/framing square that is no longer perfectly square (aka "true"), before throwing it away, try this little trick. Using a center punch, hammer a hole at the inner or outer corner of the square. If the sides of the square are too close together, punch the inner corner of the square. If they're too far apart, punch the outer corner. After a few whacks of the punch, check your progress and keep checking until your square is true again. You can find out more about this easy repair here: familyhandyman.com/article/how-to-fix-a-square.

⭐ SPLITTING A CARPENTER'S PENCIL

Maker Emory Kimbrough writes: "You can gently push the graphite lead of a carpenter's pencil a few millimeters into your running bandsaw blade to create a split lead with two points that will draw two closely spaced parallel lines. Use this to draw that freehand curve you plan to cut out with the band saw. The gap between the lines is exactly equal to the width of your saw blade, so guiding your saw right between these double lines makes for an easier job as you twist your workpiece back and forth to follow that fancy curve." **[EK]**

Carpenter's pencil

Split lead of pencil

⭐ MARKING STUDS WITH A FRAMING SQUARE

In a *See Jane Drill* video, Leah shows off a feature of a framing square that many newbie carpenters may be unaware of. Framing squares have two arms on the blade — the thicker bottom blade, called the body, and a thinner arm (90° from the body), called the tongue. The tongue is designed to be the exact width of a stud. To mark stud placement on a 2×4 base plate, you simply measure 16" along the plate (for standard 16 on-center stud placement), back off half the width of a 2×4 (which is actually a 1½"×3½"), so ¾", line up the tongue of the square as shown in the top righthand illustration, and strike a line on either side. Your stud will now be exactly centered on the 16" interval. See Leah's video for more: youtu.be/ExSZ3-bxO50. **[LB]**

⭐ MARKING IN TIGHT SPOTS

Steven Roberts, of Nomadic Research Labs, sent me this sweet little hack when I put the word out for measuring and marking tips. Steve was installing a drawer in his live-in boat to hold his digital piano. He needed to mark the keyboard's feet so that he could drill wells into the drawer to secure the instrument in place. But how to reach under there? "Easy! Just nip the end off of an old pencil, grab it with hemostats, and reach in through the gap between piano and shelf." **[SR]**

★ MAKING A COPY OF A CIRCULAR OBJECT OR OPENING

If you have a circular object or a cylindrical object with an opening, you can make a copy of that circle using masking tape. Place enough tape over the circle to completely cover it. Then, using a piece of sandpaper, sand off the excess tape around the edges until you have a perfectly round copy made of masking tape. You can then remove and transfer this template anywhere you like. This tip comes from Laura Kampf and you can see it and more great measuring and marking tips in this video: youtu.be/MiOWqjewYxw. [LK]

★ USING POST-IT NOTES TO MARK AN ANGLE

Laura Kampf offers this easy way of measuring a non-standard angle so that you can cut a perfect miter (a perfect joint made by cutting two pieces of material at an angle). Place two Post-it Notes in the inside corner of the angle to be cut. Mark the junction of the pieces where the two Post-it Notes meet. Now fold the top Post-it along those two marks. That is the angle you need to cut the two pieces on your saw to miter them together. See Laura's video for more details: youtu.be/MiOWqjewYxw. [LK]

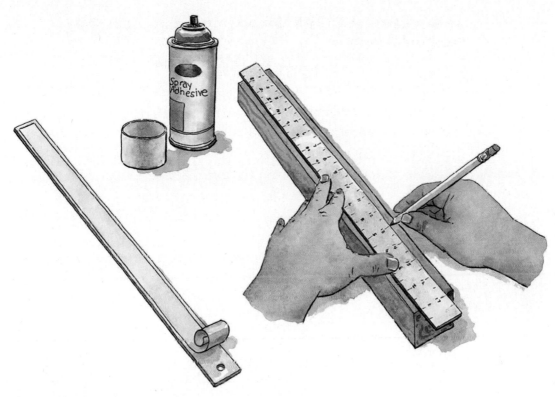

⭐ CREATING NO-SLIP RULERS

You can create no-slip rulers by spraying some 150-grit sandpaper with spray adhesive and attaching it to the backs of your shop rulers. This is especially handy for long yardstick-length rulers used for marking lumber and the like. For 12" rulers used for smaller or more fragile projects, you can get a similar effect with masking tape.

⭐ USING A LAPTOP AS A LIGHT BOX

I love this little hack from Brett of Skull & Spade. In creating art for hand-carving a rubber stamp, he laid his laptop screen-down on the table and used it as a light box to trace digital art onto paper. This is definitely a "Why didn't I think of that?" sort of tip that I will surely be using. **[BM]**

DIGITAL CALIPERS AND HOW TO USE THEM

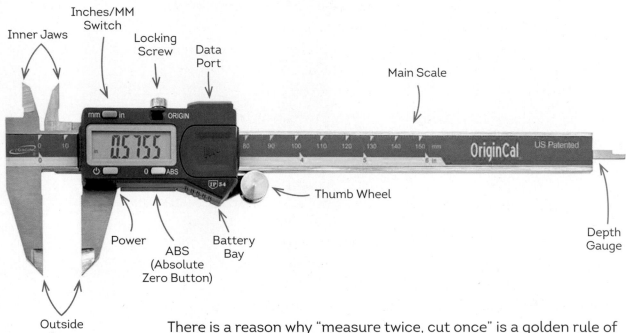

Inner Jaws

Inches/MM Switch

Locking Screw

Data Port

Main Scale

Thumb Wheel

Depth Gauge

Power

ABS (Absolute Zero Button)

Battery Bay

Outside Jaws

There is a reason why "measure twice, cut once" is a golden rule of making. Getting accurate measurements can be critical to the success of many projects. While the measuring abilities of a common Imperial/metric ruler are sufficient for many measuring situations, sometimes you need much higher precision (like when measuring parts to feed into a computer design for 3D printing or in precision metalworking). It is then that you need to reach for a set of digital calipers.

You might think that getting a set of digital calipers in your toolbox is an expensive proposition, and not worth it for the few times most of us might need such a tool. Not exactly. For under $10 at online stores and discount tool markets like Harbor Freight, you can get a worthy set of calipers that can reliably measure at a 0.0005"/ 0.01mm resolution. Many of these cheap calipers work just fine, once you learn to work around their eccentricities. If you want higher confidence in your tool, you can get a surprisingly high-quality set for under $40. My favorite are the iGaging Absolute Origins calipers. They deliver everything you need in such a tool at a really reasonable price. They even have a data-out port. With an additional special cable, you can send your measurements directly into a CAD program. Unfortunately, that cable will cost you twice as much as the calipers (but you can sometimes find them at a lower price on eBay). There are a number of projects online that show you how to make your own data cable or even allow your calipers to go wireless. If you're handy with electronics, check these out.

ALL MODELS OF DIGITAL CALIPERS ARE DESIGNED AROUND FOUR BASIC TYPES OF MEASUREMENT:

- **Outside Diameter (OD)** —The main jaws of the calipers (called the "outside jaws") are designed to measure the outside diameter of objects (usually up to 6"/150mm). Many models of calipers have a zeroing function. To do a measurement, you close the jaws, press the Zero button, and then spread the jaws to take your measurement. The results appear on the LCD display. Most caliper sets have Imperial and metric selections. Some also include a fractional measurements mode.

- **Inside Diameter (ID)** —The smaller "faces" at the top of the calipers are for measuring the inside diameter of an object. The jaws here are inverted from the main ones, but work in the same way. You open them to the walls of whatever inside diameter you are measuring and find the reading on the LCD.

- **Depth** —On the back end of a caliper's ruler/scale is found a depth gauge (or "depth bar"). This is used to measure, for instance, the depth of a hole. Simply extend the bar into the distance to be measured and read the results on the LCD display.

- **Step** —You can measure the difference between two surfaces using the outside measuring face (main jaws) of the calipers. To take a step measurement, place the outside edge of the sliding jaw against one of the edges you are measuring and the outside edge of the inner, fixed jaw against the other edge. The display will reveal this distance.

See this article on the *Make:* website for animated GIFs showing all of these measurements in action: makezine.com/2015/11/13/how-to-use-your-digital-calipers-7-tips. **[OS]**

Some models of calipers have a fine adjustment screw that allows you to move the jaws more precisely. Some also have a set screw so that you can tighten to hold a measurement. Knowing how much pressure to apply when taking a measurement takes some practice. It's often a good idea to take several measurements along an object and average them out to get closer to a reliable number.

OTHER USES AND MEASURING TRICKS:
- **Using calipers like a funky slide rule** — You can use calipers as a conversion calculator. Simply move the fine adjustment wheel (if you have one) on your calipers to the number along the scale that you wish to convert, then press the mode button to switch between millimeters, inches, and fractions (if your set includes factions). Digital slide rule!
- **Measuring the balance thickness of a "blind hole"** — Let's say you have a blind hole (a hole that does not go all of the way through) in a workpiece and you want to know the difference between the depth of that hole and bottom edge of your workpiece. Use the depth bar to measure the depth of the hole, zero out your calipers, and then measure the overall height of the piece. The result shown will be the difference between the hole and the overall height of the piece.
- **Measuring center-to-center distance** — When measuring center-to-center distance between two holes, first measure the hole diameter. Then zero out your calipers and measure from near-edge to near-edge of your two holes. (The hole diameter, which you zeroed out, will be automatically subtracted.) This gives you your center-to-center distance. This is particularly useful when trying to figure out where to place PCB mounting posts when designing project enclosures.

TIP: Always store your calipers in their case (usually hard plastic and reasonably shock-resistant). They are a precision instrument and should be treated as such. Many calipers come with an extra battery. Store it (or its replacement) in the case. Keep the faces and the scale of your calipers clean, and especially, clean the faces before each use.

TIP: Digital calipers are precise, but the world is a bit off-square. When you measure something, especially when you know what the value should be, be mindful that the measurements will likely be off by manufacturer tolerances. 9.97mm is 10mm.

- **Determining if hardware is Imperial or Metric** — If you're not sure if a piece of hardware is measured in metric or Imperial, measure a dimension of it and switch between millimeters and inches. The read-out that is closer to a whole number is likely the system of measurement used.

5

CUTTING

A friend of mine who collects knives and edged weapons once told me: "A good knife, it wants to cut you." When he said it, I thought it was kind of an odd, inscrutable statement. I was working for *National Geographic Adventure* at the time, editing a column on adventuring tech. For that column, I'd requested a handmade titanium flick knife, called a Sebenza, from celebrated South African knife-maker Chris Reeve. The gorgeous knife arrived, I took it out of the box, and, within seconds, I was bleeding. I gashed my thumb trying to one-handedly flick the stiff blade open and closed. Remembering the animistic wisdom my friend had shared, I had to laugh. Be careful. A good knife, it really does want to cut you.

⭐ CORRECTLY SET YOUR BLADE DEPTH

When cutting material on a circular saw, you want to always make sure that your saw's blade depth is set correctly. You might not think this really matters, but it does. You want your blade to only be ¼" to ½" below the bottom of the material you're cutting. This is not only safer, it helps prevent binding (where the material pinches in on the blade) and allows for a more efficient cut.

THEY ALMOST CALLED ME STUBBY

The weather had been terrible, one huge snowstorm after another. We were way behind on a four-plex build and that meant that even the general contractor (me) had to pitch in and work hard to finish.

Late into the night before Christmas, I was cutting OSB (oriented strand board) with a Skilsaw. I had two full sheets on sawhorses and was cutting along when it dawned on me that I might not have the blade set low enough. Without thinking, I reached under the OSB to see if I could feel the blade. In the fog of fatigue, feeling a spinning blade with my fingers made perfect sense. As luck would have it, I was only cutting through one of the sheets, so my fingers were safe. When I realized what I'd just done, I immediately stopped for the night. Being nicknamed "Stubby" for the rest of my life seemed too great of a price to pay.

—Jack Bonawitz

⭐ PEG SHARPENING

Elisha Albretsen from the YouTube channel *Pneumatic Addict* had some wooden pegs that didn't quite fit in their target holes. Her Revlon makeup pencil sharpener to the rescue! The sharpener was just the thing to shave a bit of material from the pegs to get them to fit. **[EA]**

⭐ PREVENTING A FLUSH-CUT SAW FROM MARRING

When using a flush-cut saw for trimming protruding wooden dowels so that they're flush with surfaces, you can prevent scratching and saw marks around the area by cutting a hole in a piece of thin cardboard and placing it around the dowel. The card acts as a shield to protect the wood while cutting the dowel flush.

⭐ PREVENTING TEAROUT WITH MASKING TAPE

Sawing through plywood, especially with a jigsaw, can create a lot of "tearout" (where pieces of the material you're cutting give way along the edge of the cut). To prevent this, cover your workpiece with masking tape around the area of the cut. As a bonus, you can draw your cuts/project layout directly onto the tape.

⭐ PREVENTING BINDING ON A MITRE SAW

If you're cutting a board with a so-called "live edge" on a mitre saw (meaning a non-factory-straight edge), to prevent binding of the blade, start the cut from the back of the board and pull forward. See the "DiResta Mitre Saw Tips" video on YouTube for more details: youtu.be/9w3Aa-mfNJA.

★ SAFELY HOLDING A SMALL PIECE OF WOOD FOR CUTTING

If you're cutting a small piece of material on a mitre saw that's too small to safely hold against the fence with your hand, use another piece of wood to hold the material you're cutting. Make sure part of the holding piece is firmly against the table's fence and that you are pressing firmly down on the workpiece you are cutting (and that your hand is a safe distance from the blade and workpiece). See the "DiResta Mitre Saw Tips" video: youtu.be/9w3Aa-mfNJA.

★ CUTTING ANGLES OF THE SAME SIDE BY FLIPPING THE WORKPIECE

Let's say you're making a picture frame and you want to cut four frame pieces with 45° mitre cuts on both ends. You can cut those two ends by making one cut at 45-degrees, moving the wood to where you want the second 45-degree cut, and then flipping your workpiece to make the cut. This way, you can make all of your 45° cuts without having to re-set the saw to 45-degrees in the opposite direction. For more, see the "DiRestaMitre Saw Tips" video: youtu.be/9w3Aa-mfNJA.

★ MAKE A TEMPORARY MARK ON YOUR SAW TABLE

If you're cutting a bunch of short pieces on a mitre saw that are the same length, mark the length of the cut right on the table of your saw. Now, you have a reference mark to line up all of your repetitive cuts. When you're done, you can wipe the mark away. **[JD]**

★ ANALOG CUTTING OF DIGITAL FILES

One thing that successful YouTube makers are often criticized for is doing too many projects that are unapproachable to their audience members who don't have a shop's worth of expensive equipment, often donated by sponsors. Maker Jon-a-Tron addressed this in an excellent Instructables piece. As he argues and demonstrates, you can cut out most digital design files using conventional hand tools. Every YouTube maker who posts projects using CNC, 3DP, laser etching, etc., should always address this in their projects, even if it's as simple as pointing out to the audience that there are many makerspaces, libraries, and other local organizations that now offer digital fabrication tools: instructables.com/id/Digital-Fabrication-By-Hand.

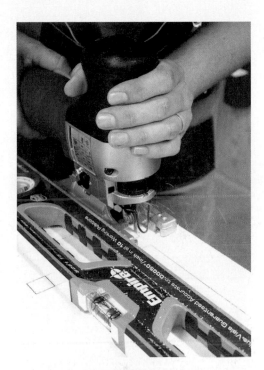

★ DIY OSCILLATING SAW BLADES

Blades for oscillating saws can be expensive. In a video on Chris Notap's YouTube channel, he explains that you can use much cheaper utility knife hook blades (which run about $10 for five) in your tool instead. All you have to do is chuck one hook end into the saw and you're good to go. **[CN]**

★ EZ-LOC CUT-OFF WHEELS

I was unaware of these Dremel EZ-Lock quick-change mandrels for rotary tools until recently (via Donald Bell's Cool Tools YouTube reviews). They come in a kit with the mandrel and a set of durable, fiberglass cut-off wheels. I hate changing the wheels that come with a Dremel using that little slotted attachment screw. And the wheels are so fragile and shatter easily. I bought an EZ-Lock set immediately. **[DB]**

YOUTHFUL ENTREPRENEURSHIP AND THE MANUAL BANDSAW HACK

My dad had a bubble letter font that he used to draw. He started drawing it when he was in high school. He would draw the words out on wood and then cut them out on a bandsaw. For letters with holes, like "O" and "B," he'd drill them out with a ¼" drill.

My dad suggested that me and my brother John go to school and get names from other kids, at like a quarter a letter. Then he would draw out all of the letters and we would cut them out on an old HomeCraft jigsaw. My brother John started it, and then I did it, when we were in elementary and middle school.

At the time, my dad wouldn't let us use the bandsaw. He thought it was too dangerous. When I was like maybe 10 my dad let me use the bandsaw. But he would cut the electricity to the shop when he wasn't around so I couldn't use the tools. But my cousin and I would go in there and take turns turning the big wheel while the other person would be in front of the blade cutting something. Manual bandsaw!

It was this that led to me learning the bandsaw and working in a sign shop, cutting out letters all through high school. I would sit there with my Sony Walkman on and the sign shop's art department would bring in piles of plastic letters with the patterns spray-glued onto the letters, and I would just sit bandsawing them all day long. Over time, I started to see the letters I had cut out up on signs around my community. Sometimes, I'd be 20-30 miles away from home with my friends, and I would look up and go, "Oh my god, I cut out the letters for that sign!" All over Long Island I'd see the letters that I had cut out.

—Jimmy DiResta

⭐ HACKSAW BLADE BENCH CUTTER

From *Wood Magazine* comes this useful idea for creating a sandpaper cutting jig for your workbench. All you need is an old hacksaw blade mounted to the edge of a workbench. You can also mark a guide for various sandpaper sizes right on the bench if you like. If you don't want it mounted on the bench itself, you can create a special cutting board outfitted with the blade edge. You can also create such a cutting edge on your workbench for cutting benchtop kraft paper. Obviously, to prevent you from slicing open your stomach (or at least your shirt) when not using the cutter, you're going to want to cover the blade when not in use. You could also mount this to a board and stash your cutter away when not in use.

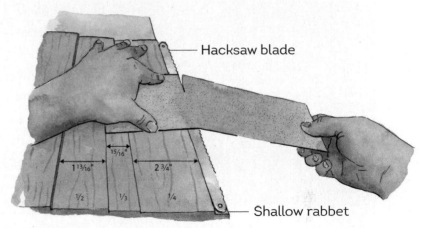

Hacksaw blade

Shallow rabbet

⭐ CUTTING PLASTIC WITH A HOT BLADE

In a recent Laura Kampf video, she reminds us that you can more easily cut plastic-based strapping, webbing, and other plastic material if you heat up your razor knife blade before each cut. Like buttah! **[LK]**

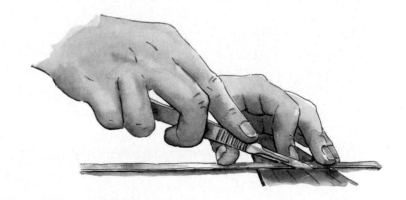

⭐ TURNING A HOBBY KNIFE BLADE INWARD

This might be an obvious tip to some, but worth pointing out in case you are unaware. To keep your hobby knife blade safer when not in use, simply turn the blade around and store it point-side in.

⭐ DON'T FORGET THE BEND ALLOWANCE

When cutting material that needs to be a precise geometry after it's been bent into a desired shape, you need to take into account the deformation and curvature of the bend. This is known as the "bend allowance." Bend allowance is impacted by the composition of the material and its thickness. You can find websites online that will allow you to calculate these allowances for different materials and thicknesses. Just search for "Bend Allowance Calculator." Find out more about bend allowance from This Old Tony: youtu.be/FyXpCPVOr8s. **[TOT]**

⭐ KNURLING YOUR KNOBS

On the YouTube channel *Pask Makes*, Neil Paskin shows how easy it is to create your own knurled knobs using little more than some rounded-off bolts, a metal file, and a vise. To knurl a knob, first you have to round off the corners of the bolt head you wish to knurl. To do this, chuck the bolt in a drill press and use a metal file on the head until it's round. Now, chuck the rounded head of the bolt and a metal file into a vise. To cut in the knurl pattern from the file into the bolt head, tap the end of the file until the bolt turns all the way around. You should end up with a decent knurl pattern on the bolt, which you can now use as a knob. Neil also took the time to build a jig that helps to better control and accelerate the process (using a power drill for turning the knurling jig). You can see that jig being built here: youtu.be/xH5qihIn5lI. **[NP]**

⭐ ROUTER BIT HOLDER AND GUIDE

On the Facebook group Shop Hacks, member Bob Commack shared this router bits storage rack that he and his young son made. They had the bright idea of edging the rack with the bits so that, at a glance, they could see the cut of each bit type. **[BC2]**

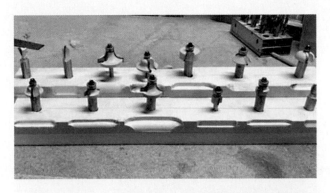

⭐ MAKING SCRAPERS FROM OLD IRON

Eugene, Oregon, artist Andreas Salzman (andreassalzman.com), takes the irons (blades) from old woodworking planes, carves his own ergonomic handles for them, and then epoxies them together. He uses them for gouging, planing, contouring, and scraping schmoo off of his workbench. **[AS2]**

Andreas Salzman

⭐ USING A WASHER AS A DEPTH GAUGE

My friend Andrew Lewis writes on Twitter: "If you are reaming out a panel hole for a switch or indicator light, use the washer from the switch as a stop on the reamer so you don't open the hole too far." **[AL]**

CLAMPING

Among the golden rules of making, up there with "measure twice, cut once" and "always clean up your shop between projects," is the idea that you can never have too many clamps. It's true. Spring clamps, bar clamps, pipe clamps, bandy clamps — there's never enough clamps. But those are only the commercial clamps. You likely already have a shop full of clamps, if you know how to see them. Binder clips, clothes pins, rubber bands, tape, old inner tubes, rope, zip ties, all sorts of things can be pressed into service to bind things together for gluing or situations where you need temporarily holding. When the need arises, don't be afraid to think outside the clamp... er... box.

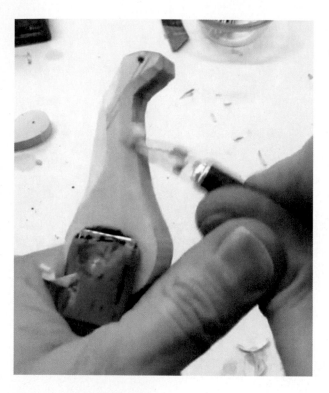

⭐ SPRING CLAMP WORK HOLDER

The amazing kinetic sculptor Tom Haney shared this tip on his Instagram page. He writes: "This is my technique for keeping my fingers cut-free. As you can see, I hold the piece of wood with a spring clamp (it keeps my fingers farther away from the blade), and I try to keep my thumbs in contact. This does two things — it gives me more power when carving, but, more importantly, it prevents my cutting hand from flying off into dangerous territory. My carving is always controlled, and my knife doesn't go far from what I'm cutting. If you can't tell, there's a lot of pressure going on between my two thumbs. Safety first, people!" **[TH]**

⭐ TURNING AN F-CLAMP INTO A WHEEL CLAMP

From the YouTube channel *Create* comes this brilliant idea. Attach a skate wheel to an F-clamp to create a wheeled clamp that you can use for moving large boards and sheet goods around. youtu.be/dZFruJITY2w

⭐ GET A GRIP

To get a better grip on your clamps that use hand-screw handles, keep a rubber jar lid grip in the shop (or several). It will allow you to apply much more tightening force. It's great to have these grips around the shop for all sorts of opening and tightening tasks.

⭐ SIMPLE CLAMP RACK

Elisha Albretsen of the YouTube channel *Pneumatic Addict* cleverly used some 2×4

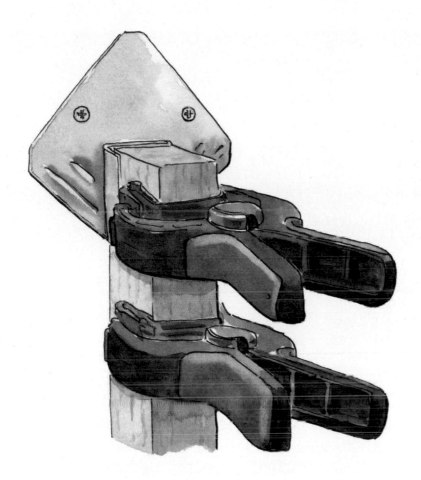

Strong-Tie clamps as the basis for her clamp storage wall. For bar clamps, she attached a short length of 2×4 to the Strong-Tie horizontally, and, for spring and bandy clamps, she attached a vertical length of 2×4. You can get these Strong-Tie clamps online and at any home store. **[EA]**

⭐ MAKING A CLAMP RACK FROM A TOWEL BAR

You can easily make a clamp rack (or "roost") using a cheap metal towel bar. You can get these in a home store for around $8 each. Of course, you could also easily make them using a wooden dowel and two blocks of wood to hold it in place. One thing about using a bar like this is that you can store the clamps with the jaws relaxed. Most of us tend to store our clamps by clamping them to a 2×4, the edge of a workbench, or similar (see the Simple Clamp Rack above). This works, but it does stress the clamp springs between use.

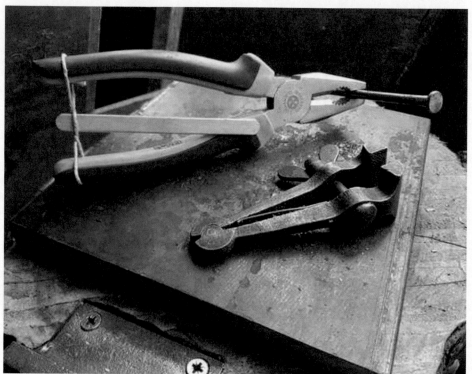

Andrew Lewis

⭐ TURNING PLIERS INTO HELPING HANDS

From maker extraordinaire Andrew Lewis comes this trick. By tightening the handle on a set of pliers with a rubber band and a popsicle stick, you can turn a set of pliers into a temporary vise for holding things.

⭐ INNER TUBE "CLAMPS"

Don't forget that you can recycle old bicycle inner tubes and use them to secure odd-shaped objects you are gluing up or otherwise need to hold together. You can wrap the tube material tightly around an object and then use a conventional spring clamp to secure the end of the inner tube to the workpiece. Also be sure to keep a complement of rubber bands of various sizes around and use them for securing smaller workpieces you are looking to clamp.

⭐ MAGNETIC JAW PADS

If you want to ensure that the steel jaws of a clamp do not scratch or otherwise mar your workpiece, you can easily create removable clamp pads. To make the pads, cut two squares of softwood, sink a shallow hole in the center of each, and glue in a rare earth magnet. Now you have protective pads you can add/remove from the jaws of your clamps as needed. You can also get adhesive-backed magnetic material and cut that to the size of your temporary jaw pads.

⭐ FELT JAW PADS FOR C-CLAMPS

The round pads on C-clamps are notorious for leaving dents in the material you are clamping. To prevent this, you can cover them with the adhesive-backed felt circles sold as table and chair leg pads.

⭐ USING CLAY TO "CLAMP" ITEMS FOR GLUING

If you have small, fragile items that you need to glue together, you can use clay, Play-Doh, poster tack, or similar materials to hold the pieces together. Glue your object together and then gently press them into the clay to keep them together while drying.

⭐ PORTABLE BENCH VISE

If you don't have enough space to permanently attach your vise to your work table, you can instead attach it to a suitably sized piece of double-thick ¾" plywood. Leave enough room around the base for attaching clamps. You can store the vise under the bench, and, when you need to use it, you can use clamps to attach it firmly to the bench.

GLUING

It is staggering to think how many different adhesive formulations there are on the market. There are gels and sprays, and there are glues formulated as thick, thin, temporary hold, and industrial strength — as well as for all manner of specific applications. Given all of this dizzying diversity in sticky-stuff, it's kind of amazing how the most common choices are still polyvinyl acetate (PVA) glue (white and carpenter's glue), cyanoacrylate (CA) glue (aka "Super Glue"), hot melt glue, and two-part epoxy. Regardless of the best glue for you, here are some tips and tricks for better sticks.

★ LABELING GLUE FLOW RATES
Many glue bottle nozzles have two or three different steps molded into the tip to indicate where to cut to establish a light, medium, or heavy glue flow (see illustration on the next page). If you use a lot of glue in the shop, you might want to have three different glue bottles with different flow rates cut into their tips.

★ FRENCH'S MUSTARD GLUE BOTTLES
Turns out, French's makes a popular glue bottle in addition to a classic yellow mustard. Many in the maker community swear by the surprisingly durable snap-top on this mustard bottle. You can also use dollar store condiment squeeze bottles and other types of recycled bottles, like dish soap containers, for glue.

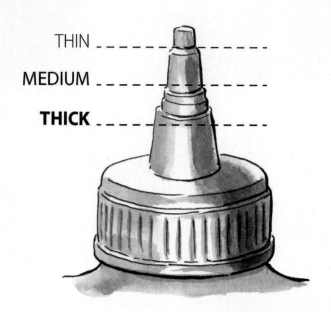

THIN - - - - - -

MEDIUM - - - - - -

THICK - - - - - -

⭐ GLUE FLOW TIP

Have you ever noticed that many glue bottles have stepped tips to their applicators? It may be obvious to some already, but worth pointing out that these are there to allow you to cut the tip so as to control the amount of adhesive you wish to apply.

⭐ REPLACEMENT NOZZLES

It is inevitable that you're going to gum up the applicator on your CA glue bottles. Packs of replacement nozzles are available for most small-bottle glue brands.

⭐ FLEXI-TIPS FOR SUPER GLUE

Maker and *Make:* contributor Jordan Bunker swears by Zap Flexi-Tips on his CA glue bottles. I've used these for years and love them, too. They afford much more precise control and allow for that capillary action voodoo that CA glue does so well. Using them helps keep the tip of the bottle from getting all gummed up. **[JB]**

⭐ LUBRICATE A CA GLUE NOZZLE

Anyone who's dealt with CA glue in small applicator bottles knows how easily the tips can become clogged. You can help prevent this by smearing some Vaseline over the tip. Also, after applying your glue, tap the bottom of the bottle on the table to knock any glue around the opening back into the bottle. This will help keep the applicator tip clear for the next use.

⭐ DECLOGGING A GLUE NEEDLE

If you use a metal needle to apply your glue, to unclog the needle the next time you use it, run the flame of a lighter or candle beneath it to remove any dried glue inside.

⭐ TINTED GLUE

There are tinted CA glues that you can use if you need to see the glue after it dries. Many formulations of cyanoacrylate come in black, brown, and other colors.

⭐ SMALL TUBES ARE ACTUALLY A GOOD IDEA

The iconic image of CA glue is in the tiny tubes branded Krazy Glue and Super Glue. These small tubes actually make sense for periodic users because CA doesn't have a great shelf life (from 30 days to 12 months depending on formulation) to ensure its full bonding strength. I've always struggled with CA glue that comes in the typical 1- and 2-ounce nozzle bottles. Even when I try and apply all of the tips and tricks suggested for good nozzle-bottle handling, mine still gets clogged and I end up risking life and limb hammering nails, pushing pins, and twisting razor knife blades into the nozzle to try and open up a clog. So, recently, I gave up the expensive brands, and I just started buying packs of "single-use" Super Glue tubes in the 2-gram size. You can get 12 of these tubes for under $7 online. And each one usually lasts for 2-3 small-job uses. Lots of makers swear by the 2-ounce bottles but if you, like me, don't like them, consider this route.

⭐ IF YOU HAVE LARGE AMOUNTS OF CA GLUE, TRANSFER IT TO SMALLER BOTTLES

Pro and "pro-sumer" brands like Bob Smith Industries, Satellite City, and others offer cyanoacrylate in a range of quantities. Two-ounce bottles are common, and you can get replacement "snip-tip" plastic nozzles for your bottle size. You can also get much larger bottles (and refill the smaller, easier-to-use bottles). You can extend the shelf life of your CA glues by keeping them in an airtight container with some silica packets. Ideal long-term storage temperature ranges between 35-40°F (2-4°C), which is another way of saying in your refrigerator.

⭐ THE RIGHT GLUE FOR YOU

CA glues come in different thicknesses, from super flowy to extra thick to gel. Which thickness you use depends on the materials you are bonding together. Thicker glues are used when you want to create a very substantive join or add a gap-filling function. Super-thin glues are good for joins where you want capillary action in which the thin glue gets sucked into the seam of your join. CA glues are also available with a variety of working times, from a few seconds up to an hour.

⭐ NOT SURE WHICH CA GLUE TO USE? START WITH MEDIUM THICKNESS

The number of different formulations of cyanoacrylate glue can be overwhelming. Start with a medium thickness glue with a 10-second (or under) working time and experiment from there for your needs.

⭐ NAIL POLISH REMOVER AS A DEBONDER

Oh no, you just glued your good side cutters together (or your hand to your forehead)! No worries. A debonder to the rescue. Nail polish remover (acetone) is a commonly available debonder. And there are commercial debonders available. You won't use this very often, so a 1-ounce bottle is all you likely need. Obviously, you want to be very careful if you're using acetone on your skin. And, please, don't glue your hand to your forehead!

⭐ TO ADD STRENGTH

Kinetic sculptor Tom Haney on using CA glue in woodcarving: "I 'saturate' softwoods with clear, water-thin CA to make the pieces harder and more durable after carving." **[TH]**

⭐ BAKING SODA AND CA TO BUILD UP MATERIAL

Baking soda and CA glue can even be used to build up enough material, in layers, to fashion a broken part. There are many examples of people using this formulation to build up material. Maker AkBKukU shows how to fix a vintage computer monitor's hinged access door on their *Tech Tangents* YouTube channel. After roughly rebuilding the missing corner of the door, the soda and CA

Hep Svadja

is filed to shape a hinge-tab that allows the little door to pivot on a plastic pin. You can see more of this example in the video: youtu.be/ n1meoZaHYZo. To see other examples, search on YouTube (or beyond) for "baking soda and CA glue." This works because the bicarbonate (baking soda) and the cyanoacrylate form an ion which reacts with a neighboring CA molecule. This starts a chain reaction (polymerization) which ends up providing not only quick setting and good adhesion, but the polymerization adds strength to the bond. **[AkBKukU]**

EPOXY COMPARISON

	FAILURE POINT		
	Gravity (lbs/kg)	Torque (inch lbs/N m)	Impact (lbs/kg)
J-B WELD EXTREME HEAT	2.5/1.13	5/.56	N/A
J-B WELD KWIKWELD	10/4.54	100/11.3	N/A
J-B WELD ORIGINAL	27.5/12.47	185/20.9	5/2.27
LOCTITE EPOXY WELD	12.5/5.67	145/16.38	5/2.27
DEVCON PLASTIC STEEL	25/11.34	200/22.6	5/2.27
GORILLA EPOXY	17.5/7.94	120/13.56	2.5/1.13

⭐ WHICH EPOXY IS BEST?

The highly recommended *Project Farm* YouTube channel does rigorous citizen scientist testing of tools, materials, lubricants, adhesives, etc. In one video, they test the point of failure of various brands of epoxy under weight, when subjected to twisting torque pressure, and surviving an impact. Under these tests, they found a clear winner: J-B Weld Original (and Devcon Plastic Steel in second).

★ CA GLUE SKIN CARE?

My friend Martin Rothfield claims that his dermatologist recommends CA glue for fingertips that have cracked open due to dry skin. Guitarists have long used it on sore fingertips caused by steel guitar strings. **[MR]**

★ WHAT GLUE IS BEST FOR XPS FOAM?

In an episode of the game crafter channel *Black Magic Craft*, Jeremy Pillipow decides to test various types of glues, adhesives, and cements to see which did the best (and cheapest) job of adhering XPS (extruded polystyrene) foam together. From his testing, Gorilla Glue Construction Adhesive and Super 77 Spray Adhesive are the clear winners. youtu.be/YCIYMVOMBso **[JP]**

★ BLOWING OUT GLUE HAIRS

Anyone who's worked with a hot glue gun knows how annoying all of those wisps of stringy glue "hair" can be. All you have to do is wait until you're done and then quickly blast your project with a blow dryer or heat gun. The hairs will vaporize.

★ STICKY SITUATIONS

You can glue almost everything with super glue, but it is not adequate for some materials. Is it possible to glue rubber to glass? Will plastic stick to wood? Once you want to join several different materials together, it can get confusing. For these moments, it's convenient to have a handy table (see page 67) of what sticks what to what. **[RH2]**

TIPS FOR HOT GLUE GUNNERS

1
SILICONE MAT
Protects
Easy cleanup
No burns

2
TEMPERATURE
High temp: high strength bond
Low temp: everyday use

3
QUICK DRY
Use can of compressed air
held upside down to freeze glue

4
MAKE ADJUSTMENTS
Use heat gun to melt bonds
for reshaping

5
BREAK THE BOND
Brush denatured alcohol around
the bond to detach

Hep Svadja

⭐ FIVE TOP TIPS FOR GLUE GUNNERS

Make:'s former photo editor, Hep Svadja, put together this little
graphic of helpful tips on using hot glue. **[HS]**

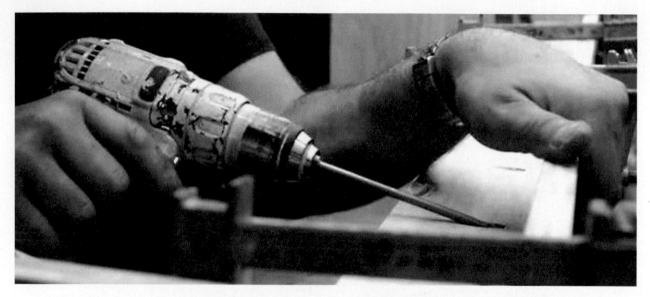

⭐ GLUING, THEN SCREWING WITH A KREG JIG

Jimmy DiResta says that when he's using a Kreg pocket hole jig to create pocket holes, as is common practice, he combines wood glue with the pocket screws. But he waits 10 minutes or so for the glue to start to set up after clamping before adding the screws. If you clamp, glue, and screw immediately, the clamping pressure against the glue can cause the workpieces to travel off-square. **[JD]**

⭐ A TAPE AND GLUE TRICK

Well-known YouTube maker Anne of All Trades posted this trick (which she got from fellow "tuber" April Wilkerson). If you want to temporarily join two workpieces, lay down some masking tape on each piece and then super glue those pieces together. This should be strong enough to hold them together (for cutting, drilling, etc.), but they can easily be pried apart and the tape removed without any impact to the workpieces. **[AAT]**

ADHESIVES CHART

MATERIALS	Paper	Fabric	Felt	Leather	Rubber	Foam	Styrofoam	Plastic	Metal	Ceramic	Glass	Balsa	Cork	Wood
WOOD	W	C/W	Sp/C	W/C/Ca	C/Ca	C	2K/H	L/Ca	2K/C/L	C/Ca	C/Ca	W	W	L/W
CORK	H/W	H/L	W	Ca/C	C/Ca	2K	W	L/Ca	C/Ca	L/Ca	Si	W	W	
BALSA	W	H/W	W	C/Ca	C/Ca	C	2K/H	C/L	2K/Ca	L/Ca	C/Ca	W		
GLASS	A/W	A	A	A/Ca	Ca	Sp	2K/Sp	C/Ca/L	2K/C	2K/C/L	2K/L			
CERAMIC	A/H	A/Ca	A/Ca	A/C/Ca	C/Ca	A	C/Ce	2K/C	2K/C/L	Ca/Ce				
METAL	A/H	A	C	C/Ca	C/Ca	C	2K/H	L/Ca/2K	2K/C					
PLASTIC	H/Sp	Sp/C	Sp/C	Sp/Ca	C/Ca	Ca	C/Ca							
STYROFOAM	Sp/C	A/H	Sp	A	L	L/A	A/Sp							
FOAM	Sp	Sp	Sp	C	C	Sp								
RUBBER	C/Ca	A/C	C	Ca	Ca									
LEATHER	F/Sp	F	2K	C/F										
FELT	A/H	F/H	H/F											
FABRIC	A/H	F/H												
PAPER	A/W													

A = All-purpose glue
F = Fabric glue
Sp = Spray adhesive
H = Hot glue
C = Contact adhesives
L = Construction adhesive (Liquid Nails, Loctite)
Ce = Ceramic glue
Si = Silicone
W – Wood glue
Ca – Cyanoacrylate (super glue)
2k = Two-component adhesive

Hep Svadja

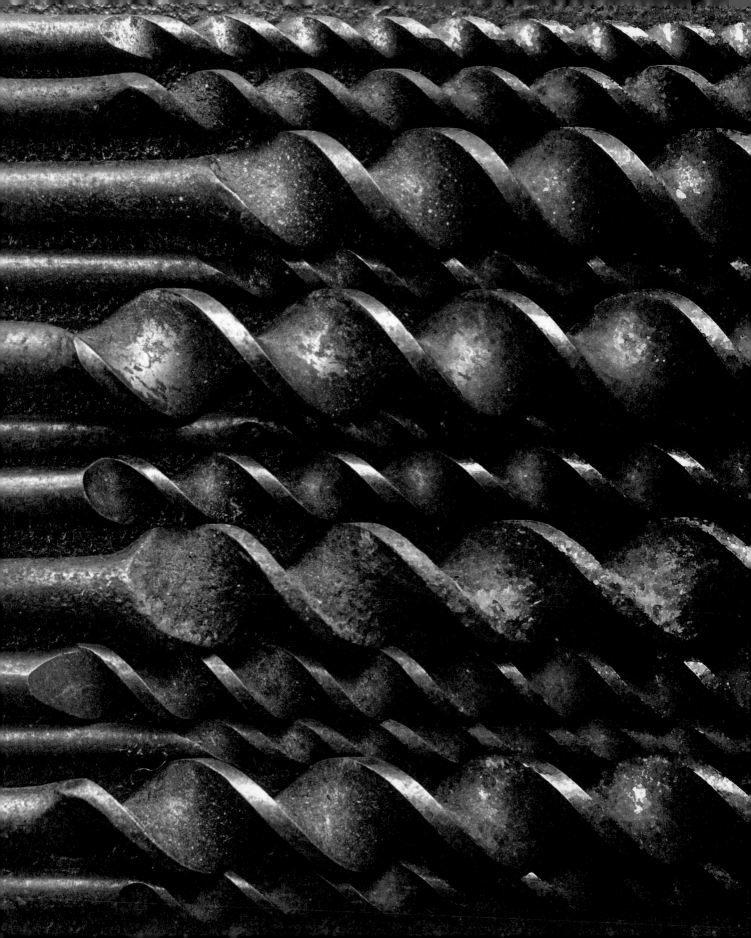

DRILLING

Jamie Hyneman of *MythBusters* fame has been quoted as saying "Making (in the physical sense) is about taking large chunks of material and making them smaller in precise ways." Along with cutting, one of the most common ways that we do this precise material reduction is through drilling.

Whatever type of making you do, from the most basic to the most complex, there's likely drilling involved. Regardless of skill level, everyone can benefit from a better understanding of drilling technology and the bright ideas that can help make the process more successful.

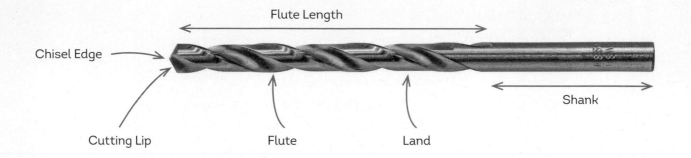

Chisel Edge

Cutting Lip

Flute Length

Flute

Land

Shank

★ HOW A TWIST DRILL WORKS

So, how does a drill actually work? This may seem like a painfully obvious question, but it actually isn't. The common twist drill is the kind of ancient engineering marvel that we completely take for granted today. But understanding what a drill bit is composed of, and why, can go a long way toward more intelligently using this most homely of tools.

A bit is a combination of several physical-science principles. It's part cutter, part chisel, part honing device, part reamer, and part material-removal system. The chisel edge and its cutting lips do most of the work, peeling away a shallow layer of material with each revolution. That debris material is then fed up along the helical flute to move it away from the business end of the drill. The flute also serves the function of getting cutting fluid to the point of the bit when such fluid is used, as in metal cutting. The portion of the drill between the flute helices is called the land. It provides the main structural integrity of the bit. It also has a secondary cutting edge. The land provides the actual cutting diameter of the bit. Finally, the shank is the main core of the bit and is used to hold it inside of a chuck.

★ USING A BIT TO DEBURR METAL TEAROUT

Matthew Perks of the wonderful YouTube channel *DIY Perks* recommends using a drill bit one size larger than the hole you just cut to manually deburr any metal tearout that happened while drilling the hole. **[MP]**

⭐ DRILLING BASICS

- Make sure you're using the correct type of bit for the material you're drilling. Bits come in many types (twist, spade, Forstner, stepped, unibit, hole saw) and have different material compositions. Look on the product packaging for what materials you can drill into or look for recommendations online for whatever you plan to cut. The main differences are bits made for cutting wood, metal, and masonry.

- Always clamp your workpiece down or place it in a vise. Drills spin. Fast. And the material being drilled would love to join the revolution. Don't let it.

- A drill is basically a spinning chisel. Chisels are happiest when they're sharp, and drill bits are, too. Sharpening drills is not as hard as you might think. There are lots of tutorials online. Keeping your bits sharp makes a big difference.

- Don't ignore the pilot. Twist drills have a tendency to travel. It's always worth the little extra effort to create a pilot hole in the center of where you want to drill. For wood, you can use a nail; for metal, a center punch.

- Start slowly as you drill to make sure you're on target and to get a feel for the material. You can speed up as you get confident. Wood can handle high-speed drilling, but for metal and masonry you want to go slower.

- The flute of a bit is designed to remove material from the hole, but the drilling process will benefit from you backing the bit out every inch or so and releasing the debris that's built up inside the flute.

- If you're drilling a throughhole, as your bit is about to break through to the other side, go slower. This helps prevent tearout on the other side.

⭐ TRICK FOR COOLING A METAL DRILL

If you need to drill a lot of holes through metal, fill a soda bottle cap (or similar tiny little cup) with whatever cutting fluid you're using as a lubricant/coolant. Position this just below the hole in your drill press table and adjust the press so that the tip of the drill bit will just dip into the cup at the bottom of its travel. Now, every time you drill a hole, you'll automatically be re-lubricating your bit. **[EK]**

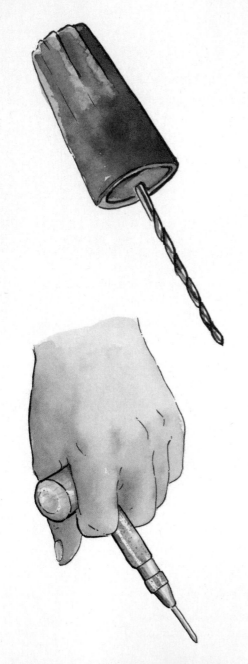

⭐ QUICK WIRE NUT TWIST DRILL

I am a big fan of pin vises, the small twist drills used in hobby modeling. On Instagram, Dennis Nestor showed off an improvised small twist drill he made by simply epoxying a drill bit into a wire nut. Yeah, that'll work. **[DN]**

⭐ ANY OIL IS BETTER THAN NO OIL

If you don't have one of the many commercial coolant/lubricants on hand that are sold as cutting or tapping fluid, then use 3-in-1 Oil, WD-40, baby oil, mineral/paraffin oil, or even cooking oil. They are all better than no cutting fluid at all. Single-weight, non-detergent motor oil can also work; however, do not use multi-weight motor oils or oils with detergents and additives. Those can corrode copper, brass, and bronze — and, even if you're not drilling into those metals, that expensive power tool you're using could have copper parts and wires inside. **[EK]**

⭐ IMPROVING YOUR CENTER PUNCH

Well-known YouTube maker Fran Blanche wanted to improve her center punch. To create a punch with more punch (allowing her to put all of her body weight behind the tool), she added a simple wooden T-handle. You can see her quick video on the build here: youtu.be/f9rEj-s_CZs. **[FB]**

⭐ CHUCK YOUR CHUCK KEY

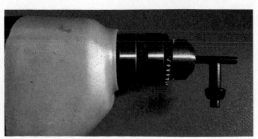

Here's a tip from the world of lathing that can be used with any machine that incorporates a chuck and requires a chuck key (which can easily get lost). You can use the chuck itself to hold your key when not in use.

⭐ WHY YOU NEED A STEPPED DRILL BIT

Adam Savage of *Tested*, Daniel McQuinn of *Real Tool Reviews*, Donald Bell of "Maker Update", and many other online DIYers are fond of telling others that they need to own at least one stepped drill bit. It's one of those tools you may not know you need, but you'll wonder how you ever got along without one as soon as it's in your toolbox. And they're not expensive, either. A basic set of such bits, with five sizes, can be had online for under $25. Adam Savage claims, in all of his years of making, he's never noticed a difference between cheap and expensive stepped bits. Here's Daniel from *Real Tool Reviews* extolling the virtues of this type of bit: youtu.be/m-jAQ23p0Hg.

Adobe Stock - Dmitriy

⭐ GET THEE A REAMER

"You've been lied to! A ¼" round drill bit will not drill a ¼" round hole. The tip of the bit will wobble around a little, even if secured properly in a good drill press. Thus, the hole will be slightly larger than the size of the bit, and it won't be exactly round. For woodworking, this tiny error is almost always too small to matter, but in metalworking it can be important. For example, you may need a shaft to turn or slide very smoothly but without any play or vibration in a hole drilled in metal. For this, you need a reamer. This tool looks somewhat like a tap, with cutting edges on the side instead of the tip. To use it, first drill a hole just slightly smaller than the desired diameter, then follow that with the reamer, taking out just a wee bit more material to create a perfectly round hole of the exact desired diameter. Yes, you'll want a drill press or a lathe for this, not a hand-held drill. An added benefit: The reamer will create a much smoother bore than a drill bit." **[EK]**

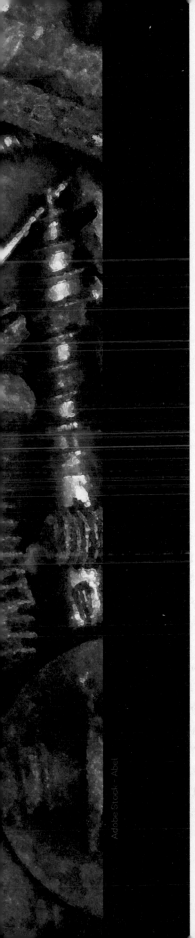

FASTENING

The word "fasten" traces back to the Old Norse *fastna*, which means to betroth or fast (in the hand-fasting, marrying sense). And, as in marriage, it is a bond that can be broken should both parties decide the connection is just not working out (although the Catholic church may not approve of this analogy). If glue is the go-to means of chemical bonding, then fasteners — nails, screws, nuts and bolts — are the mechanical equivalent.

From understanding different types of fasteners, to ways to hold them as you drive them home, to how to hide them if you want, we have tricks and techniques we think you'll find useful.

★ STAINLESS STEEL VS. STEEL BOLTS

Many people are under the impression that stainless steel bolts are superior to steel bolts. But Tom Plum of the YouTube channel *Ultimate Handyman* argues that this is not the case. Stainless steel is subject to something called "galling." This is where, after a few trips up and down the threads of a bolt, for instance, as friction (heat) is generated between the threading, a nut can seize in place. Trying to remove the nut can damage the threads of the bolt. Tom did some tests to verify how easy it is for this galling to occur with stainless steel and how it doesn't happen as much with regular steel. You also can't "cook off" seized stainless steel hardware with a torch like you can with steel. Definitely something to keep in mind when choosing hardware for a project. **[TP]**

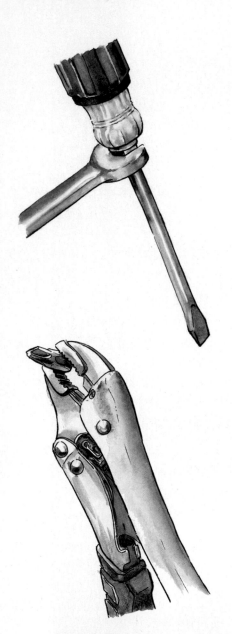

⭐ BEESWAX, NOT SOAP, AS THREAD LUBE

There is a longstanding and well-traded tip about keeping a bar of soap handy in your shop and running screw threads across it to lubricate them for easier screwing. I even included it in the first volume of *Tips and Tales from the Workshop*. But a friend of mine, Christos Liacouras, says that soap is not the best lubricant for this purpose. Soap can attract moisture and discolor your workpieces. The superior lubricant, Chris claims, is beeswax. You can get six 1-ounce bars of beeswax on Amazon for under $7 and they will last you a very long time. **[CL2]**

⭐ USING A WRENCH ON A DRIVER

Did you know that some screwdrivers have an area at the base of the handle (or part of the handle itself) that is designed to accept a wrench for greater application of torque?

⭐ SCREWING IN FASTENERS WITH AN ERASER

You can chuck a pencil into a drill and use its eraser to drive home a flat-headed screw or other fastener that is in a hard-to-reach place.

⭐ USING VISE GRIPS AS A BIT HOLDER

A pair of vise grips makes a great holder for bits if you don't have the driver handle or especially if you're trying to access a screw in a hard-to-reach place that can't accommodate a driver handle.

⭐ NAIL CLIPPERS AS AN ASSIST FOR TINY DRIVERS

Nail clippers can be handy to have around the shop. Besides keeping your nails trimmed, and removing hangnails and splinters, they have other shop uses. One off-book use for toenail clippers is as an assist for tiny screwdrivers. Because these drivers are so tiny, it can sometimes be difficult to get good torque on the driver. Nail clippers to the rescue.

THE MANY USES OF THE DIRESTA ICE PICK

Several years ago, maker extraordinaire Jimmy DiResta started selling ice picks that he was producing in his shop. This was right around the time that Jimmy was blowing up on YouTube and appearing on popular TV shows like *This Old House* and *Making It*. It wasn't long before owning a DiResta ice pick was something of a badge of honor within some maker communities.

At first, you might think: "An ice pick? What do I need that for? I don't have a big need to chip up block ice." But that was the beauty of this maker-made product (and many that he's released since on jimmydiresta.com). People immediately began finding clever uses for the tool and posting them to social media under the #direstaicepick hashtag. There are currently close to 700 items tagged this way on Instagram with people showing all of the clever ways they employ the pick, from using it to punch holes in patterns for woodworking, to hold a chalk line, as a bookbinding awl, and even as a slide for playing the electric guitar. Above is Laura Kampf showing yet another use. She's stabbed it into the leg of the bench she's building to hold a cross-brace in place while she sinks the screws.

Besides it being sort of a cool maker status symbol and a handy multi-tool, there's something very important behind what Jimmy did here. He created something slightly offbeat: a tool with an unfinished story. The user buys it and then discovers all of the ways in which it can be useful. He has continued this trend with a giant and ridiculously sharp utility knife blade, a knife blank without a handle, and other "unfinished" products that require the recipient to complete its story. Enterprising makers take note.

⭐ IS THAT A MAGNET IN YOUR POCKET?

If you're working on a project and you need a steady fastener supply at the ready, toss a disk magnet in your pocket and then stick the nails/screws on the outside of the pocket. You can do this with a shirt or a pants pocket. You can also attach magnets to your drill (or even your hammer) to hold hardware up close and personal while you work.

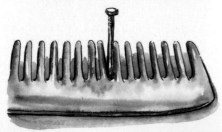

⭐ USING A COMB TO HOLD NAILS

If you have a small nail or brad that's too tiny to comfortably hold, you can wedge the nail between two tines of a comb and comfortably drive it without fear of smashing your fingers.

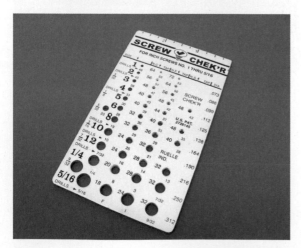

⭐ GET YE A THREAD CHECKER

Often, you will have a bin of screws and you're not sure whether they're inches or metric, what size, and whether they are coarse or fine threaded. The best way to identify these mystery screws is with a thread checker. You can get heavy-duty metal guides on Amazon, for both Imperial and metric, for under $20. One of the cool features of the Screw Chek'r brand is that it also indicates what drill bit size you need for the screw you've just ID'd.

⭐ HIDING SCREWS IN WOOD

Did you know that you can hide screws in wood by chiseling out a small flap of wood where you want the screw to go, sinking the screw beneath it, and then gluing, clamping, and finishing over the flap? Presto! I first spotted this on the YouTube channel *Average Joe's Joinery*.

⭐ PVC HINGES

On the @PneumaticAddict Instagram, Elisha shows how easy it is to create hinges in PVC by heating the end of the tubing and then using a crimping tool to flatten it out. Cut, sand, and drill a hole in the flattened area, attach hardware, and you have a serviceable hinge. **[EA]**

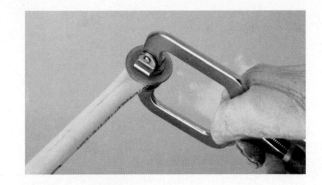

⭐ MY KINGDOM FOR A WRENCH

If you don't have a wrench the size that you need, but you have a threaded rod and a couple of nuts, you can create a funky makeshift wrench by screwing two nuts onto the rod the width of the bolt that you need to work on. Spotted on the @WhatToolsInside Instagram feed.

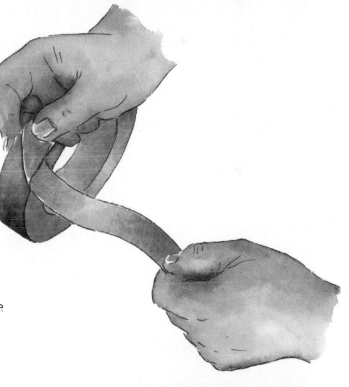

⭐ YET ANOTHER TAPE TERMINATION

There are a number of ways of terminating tape so that there's a little tab on the end to prevent the end from getting lost on the roll. Here's a way where you fold the end of the tape under itself to form a little triangle and then you hank off the end of the tape to tear the excess off.

⭐ WEAVING SPRINGS TOGETHER

On Quinn Dunki's *BlondiHacks* channel, in a video about making a tapping tool, she shares this little hack that I've never thought of. If you don't have a spring with the degree of tension you're looking for, you can create a stronger spring by simply pressing two weaker springs together. **[QD]**

10
SANDING AND FINISHING

You've labored hard planning and executing your project. You've spent countless hours with your head down: cutting, drilling, clamping, gluing, fastening. Now comes the often extra-satisfying part: sanding and finishing.

Every act of making is filled with moments of gratification and pride. The deep joy of making things by your own hands is something every DIYer knows and is likely near-addicted to. Making is like a form of therapy for many of us. And of all the processes of making, those final finishing stages (whatever forms they take) can be among the most gratifying. You're so close to completing what you've worked so hard on. These finishing steps give you an opportunity to inspect and admire the fruits of your labors. "I made this!"

Here are just a few tips to make your project wrap-up even more successful and satisfying.

⭐ CRAYON WOOD FINISH?

UK-based woodworker Matt Estlea had a thought. Finishing and polishing waxes are made out of... well... wax. And so are crayons. He wanted to know if he could finish wood with melted crayons mixed in liming wax. Basically, all he did was combine the liming wax with crushed up, melted crayons. It did, in fact, work, and the results are pretty impressive. You can see his results here: youtu.be/BAEiBxzAzbQ. **[ME]**

⭐ POPSICLE SANDING STICKS

Did you know that you can easily make sanding sticks out of wooden popsicle sticks? I just bought a set of hobby sanding sticks and I love them. But they're kind of expensive and I like the idea of making my own. This way, you have complete control over what grits you want. All you need to do is spray-glue the backs of the sandpaper you wish to use, press down popsicle sticks/tongue depressors onto the adhesive, and then trim the paper to the dimensions of your stick. You can do this to both sides of the stick with different grits, if you like. Commercial sanding sticks often have a layer of thin foam between the stick and the paper. In some situations, you might want this kind of give to your sticks. In other situations, you might want a harder sanding surface. You could easily make your own foamy sticks by adding thin foam tape between the popsicle stick and the sandpaper.

⭐ SANDING BLOCK FROM SANDING BELT

If you have old sanding belts that still have some grit on them (or extra new ones), you can repurpose them as sanding blocks by cutting scraps of wood so that the belt fits snugly around the blocks. You can use one piece of wood as a wedge to get a nice, taught fit. Jimmy DiResta shared this trick in his Instagram stories. **[JD]**

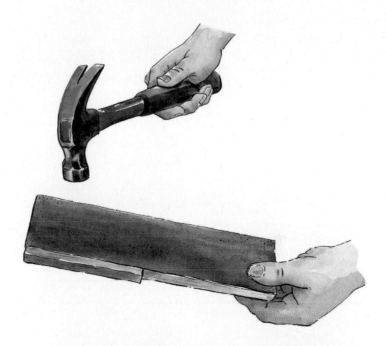

⭐ SANDING STICKS WITH HEAT-ACTIVATED SANDPAPER

You can also make sanding sticks using heat-activated 3M sandpaper. Well-known YouTube maker Bobby Duke has shown off this trick. All you do is heat the adhesive backing, burnish any size/shape sticks you wish onto it, and trim them to the size, shape of your stick. **[BD2]**

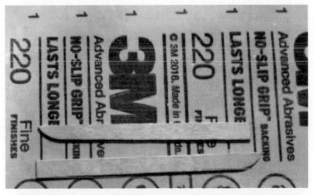

⭐ APPLYING THE RIGHT PRESSURE WHEN SANDING

Here's a great idea for making sure you're applying the right amount of pressure when using an orbital sander. Paint a red dot on the pad of the sander. As you work the tool, if you can see the dot, you know you're applying too much pressure. When the dot is nothing more than a blur, your sanding pressure is correct. **[SN]**

⭐ CREATING A SANDING MASK

After hammering rivets onto a workpiece that you want to sand smooth, you obviously don't want to mar the surface surrounding the rivets. Place a piece of masking tape over the rivet heads before filing them to protect the surrounding surface. You'll end up removing the part of the rivet that's standing proud, but the tape will remain on the surrounding surface to protect it.

⭐ SEALING KNOT HOLES WITH NAIL POLISH

Did you know that you can seal up knot holes in a wooden workpiece using top coat clear nail polish? Elisha shared this handy tip on the @PneumaticAddict Instagram page. You can also use CA glue to achieve a similar result. **[EA]**

⭐ TIPS ON FINISHING 3D PRINTS

In an installment of Bob Clagett's "Bits" series on his YouTube channel (*I Like to Make Stuff*), he ran through a number of different techniques you can use to smooth out and finish 3D prints. The techniques he recommends are vapor smoothing (exposing the print to acetone vapor in a closed container), using products like XTC-3D to add a resin coating to your print, using Bondo glazing putty, and finally, using automotive filler primer. Which technique you use depends on the project, how much of a finished look you want, what you'll be painting the parts with, and other factors. It's a good idea to experiment with all of these techniques so that you know which ones work best for you. See Bob's video here: youtu.be/ELfaQ8juSM8. **[BC]**

⭐ CUTTING A CAULKING TUBE NOZZLE

Getting a good material flow and a proper shape to your caulking bead depends a lot on how you cut the tip on the nozzle. On *Make Build Modify*, Justin Sparks has a method that he says will easily improve your ability to run a perfect bead. He makes small, successive cuts on the tip at a 45-degree angle until he exposes the hole. Then, he bevels the edges of the tip he's just cut on both sides, also at 45-degree angles. Finally, he cuts off the resulting sharp tip on the nozzle. To finish things off, he rubs the newly created nozzle in the crevice he's about to caulk to smooth out any sharp edges. You end up with a much more pointed tip than just a single 45-degree cut. See Justin lay down some sweet beads here: youtu.be/DZvoQ3kkJII. **[JS]**

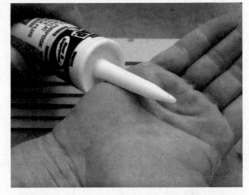

⭐ USING CA GLUE AS A FINISH

Did you know that you can use super glue (CA glue) as a finish? A little coat of super glue adds a smooth, shiny, and very durable finish to wood projects. Here's a video on how to do it: youtu.be/M-zQW4GHP9U.

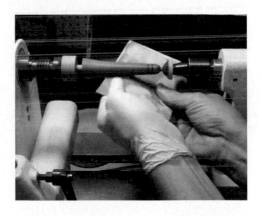

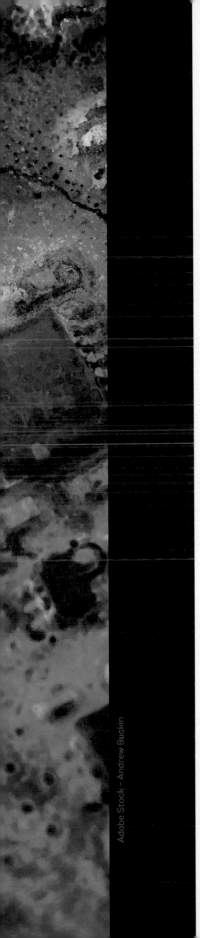

11

GENERAL ELECTRONICS

It's so heartening to see how many makers, of all different stripes, are now learning at least basic electronics. In the early years of the maker movement, I would see projects at Maker Faire or other gatherings of DIYers where people had literally cut out existing circuits from toys and other consumer electronics so that they had, for instance, an LED light or a sound effect that they wanted to incorporate in their project. Even the simplest circuit seemed insurmountable to them. One artist who incorporated video into his glass-blown sculptures came to a meeting of DC Dorkbot (the chaptered maker organization of artists and makers) with some of his pieces. He was literally embedding old portable DVD players and phones into his work because he knew of no other way to include video into his projects. One of our Dorkbot members volunteered to tutor him on how to use cheap phone screens, microcontrollers, and basic circuits to get the results he needed. It was gratifying to see how learning these skills made a real impact on this talented artist's work (and his wallet).

What follows is a collection of tips, tools, and work practices that can help both newbies and old hats alike get more from their electronics projects.

★ IT'S POSITIVE BECAUSE IT'S MORE

This might be a huge "duh" to some readers, but for those who still have a hard time remembering which leg (the longer or the shorter) on a polarized component is positive and which is negative, the Raspberry Pi Foundation tweeted this reminder: "The longer leg is + because plus means more." I always had a hard time remembering which was which until I encountered this memory device many years ago.

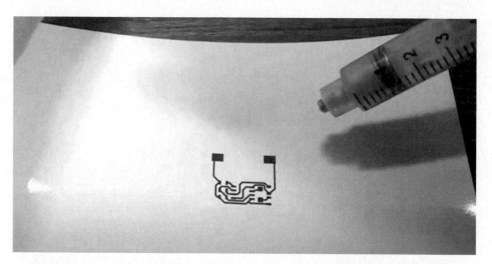

★ LASER PRINTING CIRCUIT BOARDS

There are a number of ways to create non-etched printable circuits using conductive ink and laser or inkjet printers. Maker Rich Olson uses glossy inkjet paper in a laser printer and silver conductive single-part heat-cure epoxy to print his circuits. For the epoxy, he uses Atom Adhesives AA-DUCT AD1 Silver Adhesive (which you can get on Amazon for under $20). He's experimented with various inkjet papers and recommends Epson Value Photo Paper Glossy.

The process basically involves printing your circuit on the InkJet paper, squeegeeing some silver epoxy over the image, heating the silver to adhere it to the print, and then repeating the silvering and heating steps. Once finished, you can use low-temp solder to mount components to your paper PCB. Rich offers more detailed instructions of the process on his website: nothinglabs.com/no-etch-circuit-boards-on-your-laser-printer. **[RO]**

⭐ DON'T TRUST WIRE COLOR-CODING

I have encountered two instances recently on why trusting color-coding of wires is not so wise. I had a handyman come to replace the pipes and disposal unit under my kitchen sink. He assumed the color-coding of the wires was correct and nearly fried the box, the disposal, and himself. The guy who owned the house before me fancied himself a DIYer and did some very creative renovations and repairs. Pros who come to work on my house often laugh and marvel at some of this "creativity."

I also encountered a JST-terminated wire recently where the red and white wires were switched where they entered the connector. The moral of this story: Always double-check your wires to make sure they are who they say they are. Trivia time: Did you know that JST stands for "Japan Solderless Terminal" and began life as a standard for solderless connectors in Japan?

Adobe Stock - ronst k

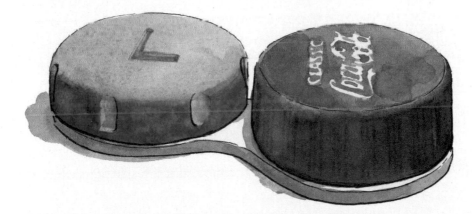

⭐ SODA CAPS AND CONTACT LENS CASES

Make: pal Miguel Valenzuela posted this idea to his Facebook page. He made a wonderful little discovery that a soda bottle cap's threads match those of a contact lens case. Actually, the discovery was made by his 2½ -year-old daughter, Charlotte (aka Charlie). Great job, Charlie! In response to the image, Lenore Edman of Evil Mad Scientist Labs reminded us that you can also use contact lens cases as tiny parts holders, especially for holding tiny surface-mount components. **[MV, LE]**

Becky Stern

⭐ OPTIONS FOR DIFFUSING LEDS

If you have done any diffusion of LED lighting in your projects, you likely already know about using things like tissue paper, frosted glass, and fabric. But there are many things you can use. Well-known maker Becky Stern has experimented with all sorts of material. Besides "the classics," she's tried underlighting, laser-cut acrylic, crinoline tubing, ping pong balls, thermoplastic, glue, and 3D-printed diffusers. Becky has an Instructable describing how she experimented with all of these methods: instructables.com/13-Ideas-for-Diffusing-LEDs. **[BS]**

⭐ MAKING A WIDER BREADBOARD

This little quick tip comes from the *Digicool Things* YouTube channel. Did you know that you can easily expand many small breadboards? The type of boards that are clear plastic with a paper backing can be cut (by slicing through the backing material) and rejoined to create the width of board you desire. For instance, you can cut and snap one of the power bus rails off one of two small boards and then rejoin the boards. The remaining power bus on one of the boards becomes a center trench that ICs can straddle, leaving much more real estate on the expanded board for hooking up your components. Amazingly, when you cut the paper backing and remove the bus, the boards just snap back together. See this little hack in action: youtu.be/TKYMAqLvCpg.

⭐ HOLDING LARGE OBJECTS WITH HELPING HANDS

Becky Stern writes on Instagram: "When a component is too large to hold with your helping third hand tool, such as in the case with this large switch, use a rubber band to hold it in place while soldering." **[BS]**

Becky Stern

⭐ FREEFORM SOLDERING LED JEWELRY

Have you ever thought about creating unique LED jewelry using freeform soldering? It's fairly easy to do. All you need is some brass rod, some surface-mount LEDs, and a coin cell battery to power your creation. Instructable member jiripraus has created a number of such LED pieces. You can follow his shapes (he even has templates) or you can create your own. His Instructable can be found here: instructables.com/LED-Jewelry.

⭐ MINTY-FRESH TIC-TAC SOLDER DISPENSER

Thingiverse user Haku3D created an .STL file for printing out a little spool that fits inside of an empty Tic Tac container and dispenses solder. I have hand-spooled solder into a Tic Tac box and that works OK, but if you have access to a 3D printer, this is a more reliable, kink- and tangle-free solution. The .STL file for 3D printing can be found here: thingiverse.com/thing:3270453. **[Haku3D]**

Haku3D

⭐ WHICH WIRE STRIPPERS ARE BEST?

On Andreas Spiess' recommended YouTube channel, he looked at six models of automatic and one pair of manual wire strippers to try and determine which are best. The tools ranged in price from $3 to $78. His conclusions? For $10 on Amazon, the FLAMEER FS-D3 strippers work perfectly fine on a wide range of wire gauges. As you might imagine, the German-engineered, high-priced KNIPEX Tools 12 42 195 MultiStrip 10 Insulation Stripper, 8-32 AWG ($99 on Amazon) handled the widest range of gauges and performed the best overall. **[AS]**

Adobe Stock - SERSOLL

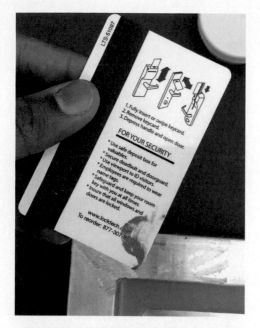

⭐ HOTEL KEY CARD SOLDER PASTE SPREADER

On Twitter, the brilliant hardware engineer Mohit Bhoite shared this little tip on yet another use for the lowly hotel room keycard. Mohit uses them for spreading solder paste for surface-mount soldering. As I pointed out in the first volume of *Tips and Tales from the Workshop,* you can also use a cut key card, or old credit card (cut on the diagonal), as a non-marring prying tool for use on the seams of plastic consumer hardware cases (or other prying applications around the shop). If you haven't seen Mohit's work, you have to check it out (bhoite.com). He has raised freeform soldering to an artform. **[MB]**

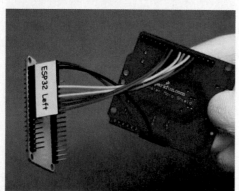

⭐ TIPS FOR WIRING UP MCU PROJECTS

DuPont connectors are a type of jumper wire connection technology that allows relatively secure electronics connection that can be plugged/unplugged. YouTuber Andreas Spiess recommends using blank Dupont connectors, custom-wired to your headers, instead of individual, pre-made connectors. This way, you can connect/disconnect the whole bank of connections without making mistakes, and you can control the wire color-coding. Another tip: If your blank Duponts don't have enough sockets on them, you can add what you need by taping additional connectors to arrive at your desired number of connectors. You can even use adhesive labels to do the joining and they can then double as an identifying label for the connector's purpose. You can buy blank Dupont connector kits online, with an assortment of header sizes, for around $10. See more DuPont connector tips in this Andreas Spiess video: youtu.be/uYf7vFREV98. **[AS]**

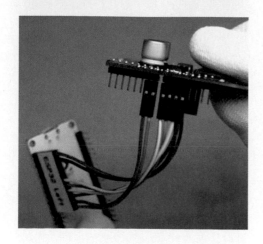

⭐ TRIMMING LEADS WITH NAIL CLIPPERS

Make: Contributing Editor Charles Platt writes: "If you trim the leads of a component using any of the usual tools, the cutting pressure flattens the lead and broadens the end of it (under high magnification). This means it will slide easily into a breadboard in the right orientation, but in the wrong orientation, it may tend to jam. My solution: Make an angled cut using nail clippers, which are much more precise than other cutting tools. A 45-degree cutting angle creates a sharp point that easily pushes into the little clip inside the board." **[CP]**

⭐ PRINTING ON HEAT-SHRINK TUBING

Did you know that you can get a special heat-shrink tubing that can go into Brother label printers and print information directly on the tubing? The only problem is, the tape costs $28 per 5-foot roll and it will only work on higher-end Brother labelers. But fear thee not. Wiley hackers have figured out that you can overcome the mechanism inside models such as the PT-H107. The PT-H107 is no longer sold by Brother but can still be found on eBay and elsewhere. These printers have a series of limit switches inside that mate with notches found in the tape cartridges. For instance, the H107 has three such switches. But, if you cover over the bottom two notches in the shrink tubing label cartridge, the tape will work. And even better, you can get knock-off shrink tubing label cartridges on websites like aliexpress.com for as little as $4 a cartridge. For more information and some links to videos on the hack, see this piece by Al Williams on Hackaday: hackaday. com/2019/08/14/print-your-own-heat-shrink-labels-for-factory-chic-wire-naming. **[AW]**

⭐ REINFORCING WIRE CONNECTORS WITH HOT GLUE

Several years ago, I made a "space vampire" Halloween costume which incorporated an Adafruit HalloWing. My friend Alberto, my son Blake, and I had fun playing around with the HalloWing, but we had a devil of a time unplugging the 2- and 3-pin JST connectors on the small, heavily populated circuit board. It's hard not to pull on the wires as you try and work the connector out. I asked Adafruit's John Edgar Park if he had any suggestions and he offered: "Hit the wire/connector intersection with hot glue to help provide a stronger connection." It worked! If you do this, make sure to go easy on the hot-melt. Since things are so tight, you don't want to add to the bulk of the connector, which might make it harder to get your fingernails in there for a decent pull. **[JEP]**

Lady Red Beacham

⭐ USING ZIP-TIES AS ACTUATORS

Maker Alex Glow wanted to add actuated wings to Archimedes, her robot owl companion. But she knew that would require more servos, more control electronics, and power. And more chances for failure. Her solution was simple. She attached the wings with plastic zip ties. Now, the wings bounce and move as Archimedes rides on Alex's shoulders. **[AG]**

Amie DD

⭐ UNDERSTANDING SERVO HORNS

Amie DD asked her Twitter followers to give her some insight into the different servo horns that usually come with a servomotor and what they are specifically used for. The collective wisdom (referenced by Amie in this image using Scooby-Doo Lego figures):

• **The Daphne Horn (6-arm horn):** Multi-purpose horn that combines the best of disk and arm types. Can serve either rotational or linear actuation functions. One commenter noted that a Daphne is what you reach for when you don't have a Fred or a Shaggy.

• **The Fred Horn (disk horn):** Used for continuous rotation or when the connection will be under significant stress.

• **The Shaggy Horn (4-arm horn):** This one was originally developed for R/C airplane control and is used for controlling ailerons because it pulls in on two sides, allowing for one side to go up and the other to go down at the same time by the same degree. The multiple holes (in all of these horns) adjust the total throw of the movement so you can set upper and lower limits. The Shaggy is commonly used in many linear control applications.

Of course, there's another wonderful tip in here — Amie's use of Scooby Doo Lego minifigs to identify parts in photos when you don't know the official names. More of this, please! **[ADD]**

12

SCULPTING, MOLDING, CASTING

When makers start talking about what new skill sets they desire to acquire, welding is often at the top of the list, right above machining. With such skills, it seems as though you could fabricate nearly anything. And these days, everyone sees desktop fabrication (3D printing, CNC routing, laser cutting) as a new frontier allowing you to fabricate your heart's desire.

That may all be true (welding is forever on my skill-set bucket list), but there is another domain of making that gives you maker superpowers, and that is molding and casting. Once skilled in these techniques, you can fabricate or duplicate all sorts of complex objects in resins, plastics, metals, waxes, and other castable materials.

There is a lot to know in making molds and successfully casting parts. The dizzying array of mold types, the eccentricities of the casting process, and many choices in molding and casting materials can be intimidating to the newbie. But no need for anxiety over learning this skill. Once you get started and learn the basics — once you get over the initial learning curve hump — you really will feel as if you have a maker superpower.

⭐ USING TINFOIL AND DUCT TAPE FOR PROTOTYPE SCULPTING

On the *Punished Props Academy* YouTube channel, prop maker Paige Cambern recommends using regular kitchen aluminum foil and duct tape to create forms that you can then sculpt on. You can squish it, build it up, and cut it down. This is a very inexpensive and surprisingly versatile material for prototyping a desired form. Once you have a shape that you like, you can cover it in foam clay or other desirable material and use it as a form for molding and eventual casting. In Paige's case, she was building a form to get the shapes and proportions right for a Spyro the Dragon hat built from EVA foam ethylene-vinyl acetate, aka poly foam) on a baseball cap. **[PC]**

⭐ SCULPTING WITH TOILET PAPER AND GLUE

One greatly underappreciated sculpting medium is toilet paper and white glue. This is commonly used in modeling for tabletop terrain boards, model railroading, and dioramas, but you can use it for all sorts of sculpting, similar to creating sculpted forms as in the foil and tape method above. In fact, you can use aluminum foil as the base material and cover it with a paper towel or toilet paper soaked in thin white glue. The only drawback to this method is that you have to wait a long time for the paper to dry. You can get best results and quicker drying times if you thin the glue 50/50 with water. You can then use the resulting form to create a mold and cast from that. One of the unsung masters of TP sculpting is DM Scotty on YouTube. Here's a baby Groot figure he made entirely from toilet paper, paper towel, and white glue (left). **[DMS]**

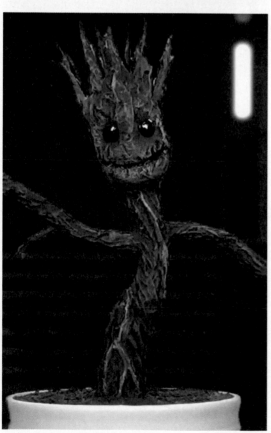

⭐ KEEPING TRACK OF PAPIER-MÂCHÉ LAYERS

Papier-mâché sculptor Jay Olson, in a video about making Halloween skulls on his channel *Unhinged Productions*, shares a useful tip for any papier-mâché work. He uses different colored paper for each layer (e.g., switching back and forth between the white and yellow pages of a phone book) so that he can make sure he has completely covered the last layer with a new one. **[JO]**

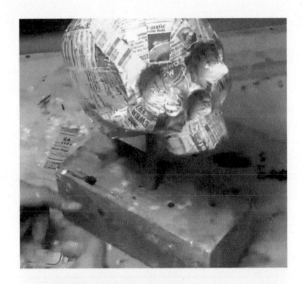

⭐ JELLO MOLDS

Did you know that you can make a quick n' dirty simple, one-off casting mold from a 50:50 mix of gelatin and glycerine? This will create a mold that is not very sturdy, but good enough for a low-cost, single-use cast where high resolution is not necessary. Puppet maker Philip Stephens shows molds that he's made using this method on his YouTube channel: youtu.be/AS7dlRPryP8. **[PS]**

⭐ MAKING A SILICONE MOLD FROM CAULK

You can create a decent silicone mold from plumbing caulk. All you have to do is squeeze the caulk into some water that has a little dishwashing liquid added to it. Then, you knead the caulk to create a dough that you can press over objects to create simple molds. The results are quite impressive. Given the high cost of silicone molding rubber, this technique may be worth a try. Epbot created these example casts (right) on their YouTube channel: youtu.be/0jFjaVuueEU.

★ ELONGATE YOUR POUR

When pouring silicone rubber into a mold, bubbles in the mold are your enemy as they will lower the fidelity of your cast. There are all sorts of ways to reduce bubbles, from shaking the mold to pricking the bubbles to putting the mold in a pressure pot to suck all of the bubbles out. One way you can help avoid bubbles in the first place is to hold the cup containing your silicone rubber as high above the mold box as possible. Don't go overboard and stand on a ladder or anything, just hold the mixture as high as is comfortable and without running the risk of missing the mold as you pour. This will help to elongate the stream of silicone and will stretch and break many of the air bubbles in the mix before the rubber hits the mold box.

★ CONVERTING A CHEAP PRESSURE POT

One tool often found in resin casting is a pressure pot, used during the molding and casting process to remove bubbles from the mold and the cast. These pots can cost upwards of $300-$400, but with a few tweaks and some additional hardware, you can turn a cheap ($99) Harbor Freight paint sprayer pressure tank (used in things like auto body painting) into a casting pot. There are numerous videos on YouTube that can show you the simple conversion process, but basically the process involves (1) removing the internal feed pipe, (2) changing the paint output hardware for a pressure safety valve, (3) adding a pressure gauge, and 4) installing a ball valve and air hose adapter to the gauge's input.

★ MOLDING FOAM WITH A SOLDERING IRON

Brett McAfee of the *Skull & Spade* YouTube channel makes fancy wooden crates for packaging some of his forged products. For the support bed upon which his products rest, he uses a soldering iron to melt designs into the foam. He then glues velvet material to hold his lovely creations. Using a soldering iron like this is common in the cosplay (costuming) and prop making communities, and other disciplines that frequently employ shaped foam. All you have to do is mark out your designs onto the foam and then use a hot soldering iron to melt those designs into the foam. You can "cut" in as deep as you like by putting as much pressure as you want. Make sure to use an iron tip other than the ones you use on electronics, as it will get gummed up by the melted foam. As you might imagine, this process will create a lot of noxious fumes, so do it in a well-ventilated space and wear a respirator. **[BM]**

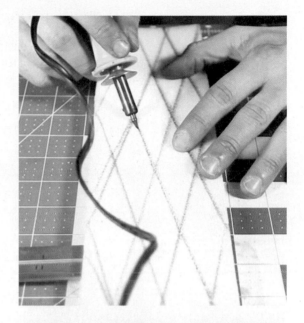

⭐ SILICONE MAKEUP BRUSHES

Well-known maker Amie DD shared this quick tip on Twitter: "I use silicone makeup brushes when applying 3D Gloop [glue for 3D printing] or resins from Smooth-On. When the material is dried on the brush, it peels right off and you can reuse them!" You can buy a set of seven Lormay silicone brushes on Amazon for around $8. **[ADD]**

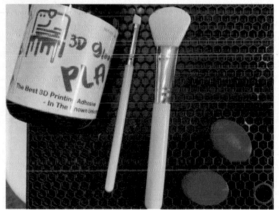

Amie DD

⭐ MAKING MOLD BOXES WITH SCRAP UHMW PLASTIC

My pal Andy Birkey sent me this great tip for building quick n' dirty mold boxes. He keeps a selection of UHMW (ultra-high molecular weight) plastic scrap pieces in his shop. To create mold boxes, he hot-glues suitable pieces of the plastic into the shape that he desires. As you can see from the photo to the right, he doesn't even bother to cut the pieces, just glues them into the desired mold box shape. The glue is strong and waterproof enough to hold the

Andy Birkey

job, but the box can be easily snapped apart after you extract the mold. Andy also writes the weights for his part A and part B mold mixes right on the box, along with what time he can de-mold (1:15 p.m., in this case). All of the writing can be wiped off with alcohol when the job is done and the UHMW can be stored away for the next time. **[AB]**

⭐ MOLD LEVELING

When pouring a casting into a mold with a lot of real estate, place the mold on a board and sink four screws into the board like table legs. By adjusting these screws, you can make sure that the casting resin is perfectly level in the mold. **[AB]**

⭐ CASTING WITH CARDBOARD AND 3D MOLDS

On the YouTube channel *XYZAidan*, young maker Aidan Leitch came up with the bright idea of 3D printing molds that he can use with cardboard pulp to create new cast paper objects. He makes the pulp in a blender by combining bits of cardboard trash and water. With this technique, you can cast all sorts of paper-based objects from 3D printed molds. You can watch his video of the process here: youtu.be/0ItPfhx3ulw. **[AL2]**

⭐ TURNING BOTTLE CAPS INTO KNURLED KNOBS

Maker Emory Kimbrough writes: "Do you need to make an easy-to-turn set-screw with a nice big knurled knob (perhaps for a jig you're building), but all you have on hand is a bolt with a tiny little head? Chug a bottle of your favorite soda and place the bottle cap mouth-up on a tabletop. Now, place the screw or bolt in the middle of the soda cap. (That's correct, the head of the screw goes against the inside of the cap.) Fill the cap with epoxy. When the epoxy sets, there's your set screw with a knurled knob." **[EK]**

BE CAREFUL YOU DON'T END UP IN CHEMICAL KOOKYTOWN

Never forget, when you're molding and, especially, casting materials, you're dealing with chemistry. It's chemistry! I have been reminded of this the hard way more than once. So, the ambient temperature, the temperature of your materials, and the depth, or mass, of your pour are all super-important because they impact the working time — that's how long you have to mix and work with the material before it heads toward chemical kookytown.

Many molding and casting chemicals are exothermic — that is, they give off heat as the chemical reaction occurs. If you ignore these factors, you will find yourself playing a very challenging game of hot potato. I once proceeded to dump the entire contents of my mix of casting material all over my work boots. Not fun. The mix had gotten hotter than I expected, and I couldn't hold onto it any longer. It had gotten super-incredibly hot. So, be mindful of these factors and keep the pour off your boots and other areas of your person, your pets, and loved ones.

When working with this material, remember that the ambient temperature in the shop is hugely important. Also, the greater the mass of the material, the faster the chemical reaction will be. If it's thin and spread out, you have more time and less heat to deal with. If it's concentrated and deep — chemical kookytown —you'll get a new surprise pair of resin-cast boots.

—Andy Birkey

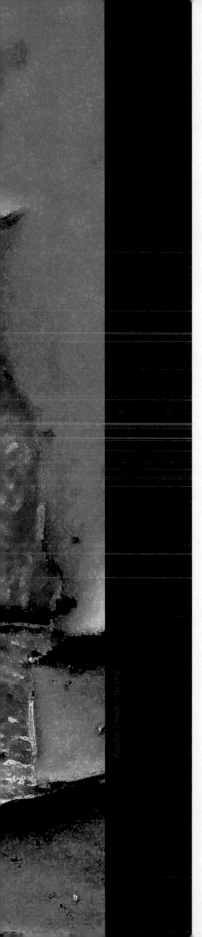

13
PAINTING

Paint is a healer. In more ways than one. Paint brings a project to completion and to life. Paint protects the object painted. Paint breathes new life into old things that have seen better days. Isn't it fun and revealing when you remove layers of old paint from the wall of a house or an old object and you catch glimpses of its past lives? Years ago, in my home, I discovered that the bedroom had, at one point, been lime green, and the kitchen an almost neon-bright yellow. (Shudder.) Which brings us to the final magic trick of paint: its ability to impact our mood, and our feelings for an object or an environment.

For many, painting is one aspect of making that they dread. As with anything else, knowing efficient and effective ways of painting can help make a daunting task far more pleasant, even fun. Here are a few tricks you can put up your sleeve for the next time you reach for a brush and a paint can.

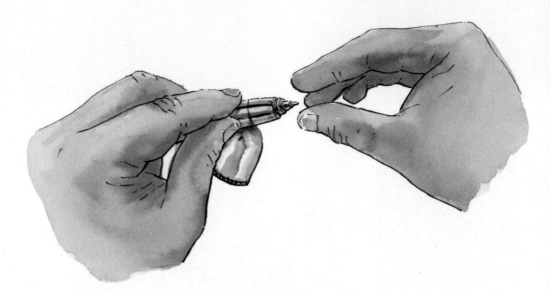

⭐ IMPROVING A CHEAP AIRBRUSH

I have a very expensive airbrush that I'm always hesitant to use because I hate cleaning it and don't want to mess it up. So, I was intrigued by the idea of buying a super-cheap brush and improving it enough to use for basic priming, base coating, and painting large pieces. In a video on Rahmi Kocaman's *Scale-a-Ton* YouTube channel, he shows how he improved the quality of a $20 brush, mainly by polishing the needle with a polishing compound. Manufacturers apparently use low-quality, roughly polished needles on cheap brushes. He claims that if you give the needle a good polish after every few uses, the brush will get better and better over time. For 20 bones, I might give this a try. He also recommends replacing the cheap O-ring with a seal made from beeswax. **[RK]**

⭐ IMPROVING CHEAP PAINT BRUSHES

In a Tested.com video, Bill Doran shares a tip I'd never heard before. We all know how great cheap craft store paint brushes can be for applying glues, doing weathering, applying faux finishes, drybrushing, and quick n' dirty paint jobs. I bought a giant bag of such brushes, of all sizes and shapes, at Michael's years ago for $7 (I still have half the bag!). But, such cheap brushes have a nasty habit of quickly shedding their bristles. That's because, unlike their more high-quality counterparts, the bristles in cheap brushes are not glued in or sewn together, they are simply crimped into the ferrule. To help keep the bristles of such brushes in place, Bill runs a bead of CA glue along the base of the bristles, right at the edge of the ferrule. **[BD]**

⭐ PAINT DRIES HALF A SHADE DARKER

Adam Savage offers this tip on paint colors: "You have to remember that colors often dry half a shade darker than they are when they're wet." Considering this while choosing, mixing, and applying paint is very important in getting the final color you desire. It took me years of painting game miniatures to realize this and to do the color "correction" in my head while choosing a color. Recently, I also saw this in action while doing some touch-up painting on a bathroom wall. I knew the paint I was applying was the paint that was already on the wall (and I knew this this "paint dries darker" rule, but it went on so much lighter, I got nervous. An hour or so later, the colors matched perfectly. **[AS3]**

⭐ MIX DARK INTO LIGHT, NOT THE OTHER WAY AROUND

Don't try and lighten a dark color by adding white or a lighter color into it. Add the darker color to the light one and bring it up to the shade you desire that way. **[AS3]**

Sara Conner Tanguay

⭐ CHEAP CLAY FOR MASKING

Besides using painter's masking tape around this door detail, for the more difficult curved areas, maker Sara Conner Tanguay used cheap modeling clay. You can get such clay at any dollar store. **[SCT]**

MY HANDS ARE COVERED IN PAINT, I CAN'T OPEN THE DOOR TO MY HOUSE!

I was shooting hoops in my parents' driveway years ago when our next door neighbor appeared, his hands covered in paint. "Do you have some turpentine I can clean my hands with? I can't use the door knob to get into my house!" I asked him, "Is it oil-based paint?" He paused, and said, "Yes." I said, "Okay, you need olive oil." I opened the kitchen door and called in to Mom, who soon appeared with the Green Jug. (Like all Greek mothers, she always decanted the store-bought olive oil into a thick green glass jug for better preservation).

The next door neighbor cautiously held out his hands as Mom poured olive oil into them. He rubbed, and to his amazement, the paint dissolved easily. As he used a paper towel to clean off his hands, he looked at me and Mom like we were from another tribe, keepers of Ancient Ways.

—Christos Liacouras

⭐ USING THE BACK OF A RAZOR KNIFE

My friend Andy Birkey, who does a lot of Gothic and church restoration and repair, shared this great tip. He uses the back edge of an X-acto knife or scalpel blade to burnish the edges of masking tape when he's painting. Sealing the tape well against the surface of the object prevents paint from seeping underneath and ruining a perfect painter's edge. [AB]

⭐ KEEP DUST AWAY WHEN USING GLOSS OR CHROME PAINT

If you need to use a gloss or chrome paint for a project, any dust in the finished paint job will stand out like a badger on a billiard table. If you don't have a dedicated spray booth, then you can minimize dust by lightly spritzing the whole area with water before you start painting. [AL]

⭐ PAINTING DON'TS

- **Don't** be tempted to buy cheap paint (unless it's quality paint on sale). Don't buy cheap rollers or brushes, either.

- **Don't** skip the step of cleaning walls (or really anything other than wood) before you paint.

- **Don't** skip the step of sanding walls (and really anything else) before painting. Sanding creates a smooth but also micro-toothy finish for the paint to stick to.

- **Don't** skip the step of power-washing the outside of your house before painting it.

- **Don't** try and paint (again, pretty much anything) in one coat. Take your time. The multi-passes will serve to create a smooth, streak-free finish. Two or more passes are your friends.

- **Don't** paint over paint that is not completely dry. Again, patience, grasshopper. A layer of wet paint over partially dry paint can cause adhesion problems and premature cracking and flaking.

- **Don't** paint directly from the can. You will introduce dust and other impurities into your can. Pour it into a tray or a special painting can that you can replenish as needed.

- **Don't** use the same brushes for regular and metallic paints. Metallics have flakes in them that you don't want in your other paints. If using very small brushes, the flakes will also accumulate near the ferrule and ruin the shape of the brush.

- **Don't** be tempted to skip the step of masking off areas you don't want painted (e.g., trim). Even if you have the steadiest hand, you will likely make a mistake or two and need to touch up your boo-boos.

- **Don't** forget to always mix your paint thoroughly. It's worth the extra effort. You can also get a paint stirrer that attaches to your drill if you don't want to tire your hands.

14

DESKTOP FABRICATION

Every year that goes by, desktop fabrication technologies — 3D printers, CNC routers, laser cutters — become more affordable and more widely available. If you don't have such a machine, your friend or neighbor, your local library or makerspace, likely does. And with advances like affordable, high-resolution resin printers, metallic, high-temperature resistant, and flexible filaments for 3D printers, the versatility of what you can make with such a device continues to increase.

And, as a technology grows, so do the clever tips, techniques, and workarounds. Here are a few that I've recently come across.

⭐ CURE YOUR NAIL POLISH AND YOUR SLA PRINTS

I've written in the past about the DIY wonders that await you at beauty salon supply stores (or that section of Amazon). You can get super-cheap makeup brushes to use in painting (especially drybrushing), inexpensive, tiered acrylic shelving for storing small paint bottles and other small supplies on a work desk, and salon creme bleach for restoring faded plastic tech enclosures (see Chapter 16, "Restoration"). Add to this list resin curing. For my small print-area resin printer, I bought a UV nail curing station on Amazon for $20. It's perfect for curing small printed parts and gaming miniatures. And my wife likes to use it to cure her nail polish.

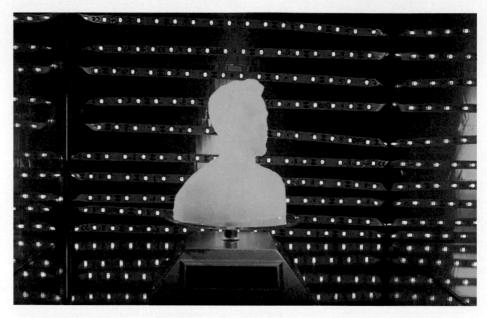

⭐ BUILDING A RESIN CURING CABINET

Affordable SLA (stereolithography) 3D printers are finally finding their way into the consumer market. Unlike their older siblings, FDM (fused deposition modeling) printers, resin-based SLAs require curing of the prints under UV light. You can buy small UV light boxes for this purpose, but you can also very easily and cheaply make your own. Bob Clagett of *I Like to Make Stuff* built his own resin printing finish station. He built a simple wooden cabinet to hold the printer and the printing supplies. Next to that is a curing cabinet outfitted with mirrors and UV LED strip-lights. As Bob points out, you could even build a curing cabinet out of a cardboard box, reflective tape, and a strip of UV lights. You can get 20' of AC-powered UV strip lights for under $20 online. **[BC]**

⭐ TESTING TOLERANCES IN 3DP

Maker Jonathan Whitaker shared this tip to Donald Bell's *Maker Update* YouTube series. Before you 3D print a large mechanism, or part of one, it's worth making sure that everything will fit the way you expect. For key parts like components that press-fit, or parts that need to turn, he prints off just that region of the mechanism and tests it out. It takes 10 minutes, and saves him having to re-do a much longer, larger print. **[JW]**

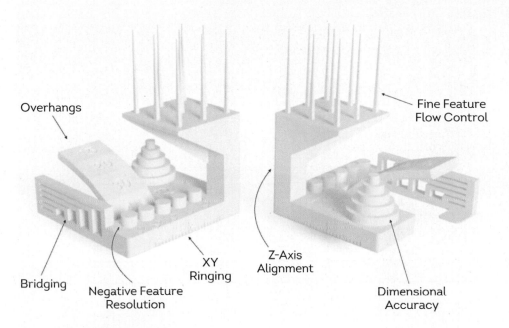

Overhangs

Fine Feature
Flow Control

Bridging

XY
Ringing

Z-Axis
Alignment

Negative Feature
Resolution

Dimensional
Accuracy

★ 3D TEST PRINT

Autodesk and Kickstarter collaborated to create a standardized 3DP test print. The idea was to create a common benchmarking print so that consumers can better evaluate and compare quality across different printers. The resulting object is designed to highlight all of the strengths and weaknesses of the printer you're testing. You can download the test print file here: github.com/kickstarter/kickstarter-autodesk-3d.

★ 50% IN-FILL IS YOUR FRIEND

Whenever you're printing small, detailed objects, avoid the temptation to create a solid print. You may think that, at a smaller size, a solid print would be better, but a solid print is more likely to have issues with "contraction forces" as the mass of solid plastic cools (pulling in some of the outer layers and distorting details). In addition, 50% in-fill cuts down on kinetic energy issues if you drop the printed object, making for a more resilient piece. See the video "How to 3D print miniatures on a FDM printer" on *The Tomb of 3D Printed Horrors* channel for more tips on 3D printing small objects, in this case, fantasy miniatures (youtu.be/AqEWl51s9Rw).

ALWAYS MOUNT A SCRATCH MONKEY*

I built a 3D scanner as part of my PhD. The first time I tested it, I was alone in my office, along with some 2,000-year-old ceramic relics. The 3D scanning process is entirely passive —a laser bounces off an object on a turntable and is observed by a camera connected to a computer. I thought I'd test it on one of the ceramic fragments, but in the last few seconds changed my mind and switched to a resin model of Columbia from *The Rocky Horror Picture Show* instead.

I clicked the scan button, the laser turned on, and the turntable spun around at about 200 RPM, firing poor Columbia clean across the room. I'd mixed up some of the control code for the motor. If I'd used the real ceramic fragment, I'd have been having a very awkward conversation with the museum curator.

The takeaway from this story is that you can never check your code enough, and you should always use a crash test dummy first.

–Andrew Lewis

Editor's note: The venerable Hacker's Dictionary *(1991), based on the much earlier online* Jargon File, *tells the origins of the phrase "Always mount a scratch monkey":*

"This term preserves the memory of Mabel, the Swimming Wonder Monkey, star of a biological research program at the University of Toronto. Mabel was not (so the legend goes) your ordinary monkey; the university had spent years teaching her how to swim, breathing through a regulator, in order to study the effects of different gas mixtures on her physiology. Mabel suffered an untimely demise one day when a DEC field engineer, troubleshooting a crash on the program's VAX, inadvertently interfered with some custom hardware that was wired to Mabel."

From there, "mount a scratch monkey" became a cautionary tale of never committing resources to a project you can't afford to lose if something goes wrong, and to test your project first in ways that won't destroy it (or key components) if something goes awry. In Andrew's case, Columbia was his scratch monkey.

⭐ UNDERSTANDING EXTRUSION WIDTH

In a video on CNC Kitchen called "Extrusion Width — The Magic Parameter for Strong 3D Prints?" (youtu.be/9YaJ0wSKKHA), Stefan Hermann looks at extrusion width, something that most users of 3D printers usually ignore. "Extrusion width defines how

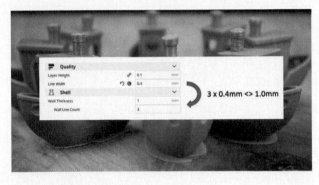

wide the lines of extruded material are," he says. "Higher values require more pressure to squeeze the molten filament out of the nozzle that also presses the layers together. I investigated how different extrusion widths impact quality and strength of our 3D prints and discovered that this might be a way to produce strong prints in a short amount of time." **[SH2]**

⭐ UNDERSTANDING TOPOLOGICAL OPTIMIZATION

Have you heard of something called "topological optimization?" It's basically a method of allowing a computer design program, such as Fusion 360, to create an optimal design geometry for an object that you wish to create. You tell the program what the use requirements of your object

are (e.g., its loads) and the computer figures out the optimal design for that object. In a video on his channel, called "Making Strong Shelves with Topological Optimization," Thomas Sanladerer explains the concept and shows you a set of shelf brackets he designed and 3D printed using this method. BTW, topological optimization is available in the free version of Fusion 360. See Thomas' video for more info: youtu.be/3smr5CEdksc. **[TS2]**

⭐ THREADED INSERTS FOR 3D PRINTS

On the *CNC Kitchen* YouTube channel, Stefan conducted a fascinating set of experiments to find out how much better metal threaded inserts are over 3D-printed threads that you might build into a 3D print. The conclusions: If your connection is going to

receive a lot of stress and/or disconnecting/connecting, inserts are the way to go. If not, you can probably get away with threads designed into your prints. The other advantage of inserts? The satisfying act of sinking them into the print by heating them with a soldering iron until the plastic melts. **[SH]**

⭐ FILL HOLLOW PRINTS WITH PLASTER

A useful tip from Zac, who makes under the name of Gimme Builds: "Here's a trick I use when vacuum-forming hollow prints. Fill them with plaster of Paris. It mixes up quickly and is hard enough to support by morning if you let it sit overnight. It also gives them weight so that they don't shift on the vacuum bed." **[GB]**

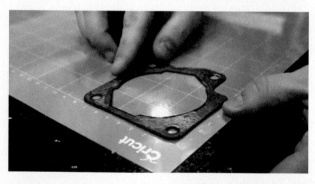

⭐ CUTTING GASKETS WITH A CRICUT CRAFT CUTTER

On *The Good of the Land* YouTube channel, homesteaders Justin and Amy show how they figured out how to use a Cricut craft cutter to create near-perfect rubber gaskets. The process involves little more than taking a photo of your existing gasket, opening it in a graphics program, cleaning it up and scaling it properly (the only challenge here), and then sending it to the Cricut. Load your gasket material and print. See their video here: youtu.be/hejUzvfGuY8.

Winston Moy

⭐ LOW PROFILE VISE FOR WORKHOLDING

Winston Moy writes in *Make:* "Holding your material securely is a prerequisite for success in CNC, and tiny projects can be difficult to get a grip on. A low-profile vise is indispensable where you may not have a lot of z-clearance. It keeps the top face of your part exposed. However, if your part is oddly shaped, or you need complete access to the sides, then consider a different method." See more on *Make:* makezine.com/2019/01/30/micro-milling-tips-for-smaller-cnc-projects. **[WM]**

G-CODE REFERENCE

G-Code	Parameters	Command
G0	axes	Rapid traverse
G1	axes, F	Straight feed
G4	P	Dwell
G20		Inch unit mode
G21		Millimeter unit mode
G28.3	axes	Select absolute position
G53		Select absolute coordinates
G54–G59		Select coordinate system 1-6
G90		Absolute positioning mode
G91		Incremental positioning mode

G-Code	Parameters	Command
M3	S	Spindle on CW
M4	S	Spindle on CCW
M5		Spindle off
M6		Tool change
M8		Coolant on
M9		Coolant off

G-Code	Parameters	Command
F	Feed rate	Specify feed rate
S	RPM	Set spindle feed
N	Line number	Label G-code block
P	Seconds	Specify dwell time

NUT SIZING CHART

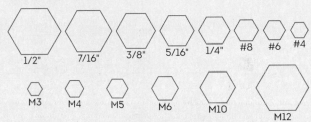

1/2"   7/16"   3/8"   5/16"   1/4"   #8   #6   #4

M3   M4   M5   M6   M10   M12

SURFACE FEET PER MINUTE

Material	HSS	Carbide
Aluminum	600	800
Brass	175	175
Delrin	400	800
Polycarbonate	300	500
Stainless Steel (303)	80	300
Steel	70	350

COMMON CALCULATIONS

Surface Feet per min =
Revolutions per min * .262 * Tool Diameter

Revolutions per min =
Surface Feet per min * 3.82 / Tool Diameter

Feed Rate (inches per min) =
Revolutions per min * Chip Load per tooth * Number of Flutes

Chip Load per tooth =
Inches per min * Chip Load per tooth * Number of Flutes

Millimeter to inches = Multiply length by 25.4

Inches to millimeter = Divide length by 25.4

⭐ CNC REFERENCE CHART

Bantam Tools has put together a handy chart that should be of interest to anyone working with CNC machines. The chart includes such useful information as metal machinability vs. hardness, plastic hardness vs. plastic melting points, common calculations used for CNC, a G-Code reference table, nut sizes, a glossary of CNC terms, and more. It's free when you join their Facebook group: fb.bantamtools.com.

⭐ GLUE-ON CARDBOARD BACKING FOR CNC

Via Jimmy DiResta comes this handy trick for setting up a CNC workpiece so that you don't have to tab the internal waste pieces before cutting. Jimmy spray-glues a piece of corrugated cardboard to the back of the workpiece. This holds everything, including the waste, in place during cutting, and then it can all be easily peeled away when finished. **[JD]**

Winston Moy

⭐ USING DOUBLE-SIDED TAPE AND OTHER ADHESIVES

Adhesives are great for holding thin stock that has a lot of surface area. You can use double-sided tape to secure PCB blanks, plastic, or plywood. **[WM]**

⭐ SECURING SMALL PIECES WITH WAX OR GLUE

"Fixturing wax or hot glue can be a good way to secure small pieces to your table. Melt it to the surface, press your stock into the molten puddle, and maintain pressure while it cools. A nice benefit with these methods is that there's no sticky residue to gum up your end mill once you cut through your part, since wax and hot glue aren't tacky when cooled." **[WM]**

Winston Moy

★ MANAGING CURVED OR IRREGULAR SURFACES

When machining parts with curved or irregular surfaces, it can be useful to use "soft jaws," which are custom fixturing solutions with clamp faces that have the negative profile of your part machined into them. You can also cut a shallow pocket in the spoilboard to hold your parts. **[WM]**

★ UNDERSTANDING ALUMINUM

Winston Moy: "Small projects are a great opportunity to start working with metals. The most common non-ferrous metal to machine is aluminum, but not all aluminum is made equal. Different alloys and tempers can drastically affect how well it machines. Pure aluminum is relatively soft and almost acts like clay at the microscopic level — and it clogs end mills easily.

"7075 aluminum, which is alloyed primarily with zinc for strength, is a harder alloy that shears much more cleanly than pure aluminum; it 'forms a chip' as they say in the machining world. A cheaper alternative that's slightly softer but still very CNC-friendly is 6061 aluminum. 6061 is alloyed primarily with magnesium and silicon, and is a common general-purpose grade aluminum. Using the wrong alloy can cause you endless headaches, so do some research before buying. McMaster-Carr's website (mcmaster.com) is a good reference." **[WM]**

⭐ MILLING PLASTICS

With plastics, it's important to have a good ratio of feed rate to spindle RPM. Wood will char if your cutter loiters too long in a single spot, but plastic will melt, clog your end mill, and ruin your project. Heed the recommendation of your CNC manufacturer, or better yet, the cutting parameters of your tool vendor. **[WM]**

⭐ FINISHING PASSES

Winston Moy: "Finish passes are essential for accuracy and surface finish. A duplicated cut will clean up excess material that your cutter might have missed on the first pass due to vibrations or other physical phenomena. A quick way to implement finish passes in any CAM software, basic or advanced, is by simply duplicating an existing toolpath." **[WM]**

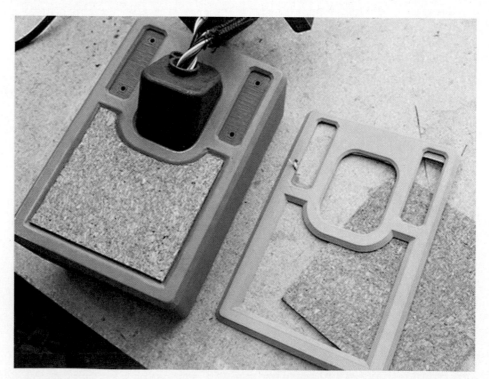

⭐ CREATING A 3D CUTTING STENCIL

Donald Bell shared this on Instagram: "A little epiphany I just had. To create a stencil for cutting this cork mat for a 3D-printed project, I printed the first few layers to get the exact outline. I glued it down with a little spray adhesive." **[DB]**

⭐ LASER ETCHING ON GLASS

Laser etching glass has a reputation for being difficult. According to Mike Clarke on his YouTube channel, it doesn't have to be. The trick, he says, is using vinyl transfer mask (also known as transfer tape). You can get this online or at any vendor that sells sign-making supplies. You simply apply the mask to the glass where your image will be and let the laser etch through the tape. This will result in much finer details in the etch. See Mike's video for more details: youtu.be/ Y5v9pBVopFU. **[MC2]**

⭐ USING LEGO BRICKS IN LASER ETCHING

When laser etching odd-shaped objects, try using Lego plates and bricks to create support structures to keep your workpiece(s) level. [Image: trotechlaser.com]

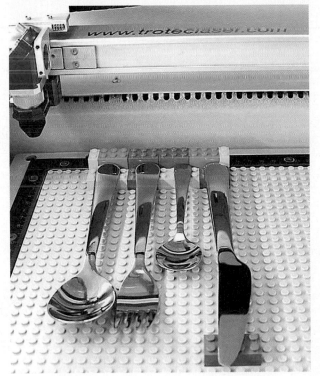

trotechlaser.com

15

WELDING AND FORGING

There are a number of skills that makers I talk to often say they want to learn. In the first *Tips and Tales from the Workshop* collection, we talked about soldering and sewing, two skills that every DIYer probably should have in their bag of tricks. The other two oft-desired skills are welding and forging, or blacksmithing.

Welding gives you the almost superhuman ability to shape metal into just about anything you want. But many people perceive it to be too hard to learn, too expensive to casually get into, and too dangerous. It doesn't have to be any of these things. There are many ways of getting into welding and to do so casually and inexpensively. You can start with little more than a home store propane torch and some brazing rod. There are lots of great videos and online tutorials on getting started in welding. I recommend the ones on Scott Wadsworth's *Essential Craftsman* YouTube channel.

Forging is an ancient and essential means of shaping metal into useful and decorative objects. Like welding, newbies can find it intimidating and think there's a long and expensive learning curve. Also like welding, this doesn't have to be the case. You can even use that propane torch you bought for baby-step welding to get started in forging. Knifemaking is often the entry into forging for a lot of makers. For that, you need little more than the torch, an anvil, a hammer, and a grinder. The parser drill bit included later on in this chapter was forged with these basic tools. There are dozens of beginner knifemaking and other blacksmithing/forging tutorials on YouTube. My friend John Graziano (*Graz Makes* on YouTube) is a knifemaker, and he recommends bladesmith Walter Sorrell's and the *Simple Little Life* YouTube channels.

Hep Svadja

⭐ UNDERSTANDING DIFFERENT WELDING PROCESSES

There are a dizzying number of techniques and technologies for welding. Figuring out which is which is the first thing that confuses newcomers. The three most common forms of arc welding are stick, MIG, and TIG. I've put together this chart (right) to help you better understand the technologies behind them and the basic pros and cons of each.

⭐ MIND THE GAP!

As you weld, metal from the stick or wire will build up at the join. If you're doing a butt joint, leave a small gap between the workpieces. This will allow someplace for the molten metal to pool and will prevent too much build-up that will have to be ground down after. A good rule of thumb is for the gap to be about the same width as the electrode/wire you're using.

WELDING PROCESSES

	STICK WELDING (AKA Shielded Metal Arc Welding)	MIG WELDING (AKA Gas Metal Arc Welding)	TIG WELDING (AKA Gas Tungsten Arc Welding)
HOW IT WORKS	A consumable flux-coated rod electrode lays down the weld. Both the electrode and the workpiece melt, forming a metal pool. This molten metal then cools to form a strong joint.	A tool called a spool gun feeds a spool of wire electrode material. The gun emits a shielding gas around the wire as it lays down the weld. This gas protects the weld from atmospheric gases such as nitrogen and oxygen, which can cause serious problems if they come into contact with the electrode, the arc, or the welding metal.	Makes use of a non-consumable tungsten electrode. The operator feeds a filler material into the weld with one hand while operating the welding torch with the other hand. A foot-based control pedal is used that dictates the heat input while laying down the weld.
PROS	• Fairly simple and inexpensive to get into. • Works in outdoor and windy conditions. • Can work on rusty, dirty metal.	• Easiest process to learn. • Higher work speeds and longer welds possible. • Neat welds with little-to-no post weld cleaning.	• Provides very high-quality and precise welds. • The most aesthetic of the three methods.
CONS	• Requires the most post-weld sanding and cleaning. • Results are not as neat as other methods.	• Workpiece must be clean and free of rust and paint. • More difficult to do outdoors than stick.	• The hardest of the three to learn and to master. • Slower process overall. • Work surface must be scrupulously clean.
NOTES	The most common and versatile form of welding. Generally the least expensive of the three. Also commonly called "arc welding." Great for beginners and DIYers. Start here.	A largely "point and shoot" process that only takes a few weeks to master. The hardest part is figuring out what gas mixture and wire to use with what material.	Great for making metal art, ornamental designs, working with stainless steel, and for automotive work.

⭐ WELDING SAFETY

- It goes without saying that welding is dangerous and flammable. You should always wear the appropriate PPE (insulated leather gloves, welding helmet, long-sleeve shirt and pants, ideally leather). Wear a welding cap under your helmet. The helmet not only protects you from sparks, but also from welder's flash. The clothing also not only protects from sparks, but also the rays of the welding arc which can burn your skin like a sunburn.
- Inspect the area where you'll be working and make sure there is nothing combustible near you. Make sure all flying sparks will be far away from any flammable or explosive materials.
- Don't touch the electrode (wire) with your bare skin, and make sure you and your clothing stay dry.
- Always make sure you maintain proper electrical grounding while you're working.
- Weld in a well-ventilated area, even if you're using a fume extractor.
- Always weld on bare metal. Breathing in the fumes from zinc and other galvanized coatings can be extremely dangerous to your health.

⭐ WELD VERTICALLY WHENEVER POSSIBLE

If the situation allows, orient your workpiece so that you can weld vertically from the top down. Obviously welding is a process of turning metal into a molten liquid. Being able to pull down and to work with gravity results in a smoother weld bead.

⭐ USING TAPE AS A THREAD MASK BEFORE WELDING

If you're welding anywhere near a bolt or threaded rod, you need to make sure you don't get any "welding boogers" (technical term, aka welding drips) on the threads. One easy way of covering this hardware is to use painter's tape.

⭐ FREE PIPE FITTING CALCULATOR

The Pipe Fitter Calculator is an invaluable tool for anyone doing welding, plumbing, or fabricating with plumbing pipe that needs to be cut and fit at various configurations and connections. This suite of apps includes calculators for things like figuring out pipe offsets (changes in the direction of piping), calculating miter cuts in pipe, curved pipe cuts, and more. You can download their collection of apps for Android or iPhone here: pipefittercalculator.com.

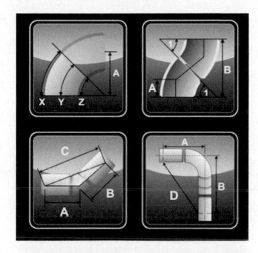

⭐ PLUG WELDING HARDWARE IN PLACE FOR A CLEANER LOOK

If you need attachment hardware on a metal workpiece and don't want that hardware (e.g. bolts) going completely through the piece, you can plug weld (a round weld inside of a hole) the bolt so that it's flush with one side of the piece. Make a hole big enough for your hardware. Tack the weld in place and then lay down a spiraling weld bead from the outside of the plug hole toward its center. Grind smooth, and when the piece is painted, powder coated, etc., you'll have a nice clean look to it.

To create more of a surface for your weld, bevel the end of the hardware first (right). This will create a larger well around the hole to accept molten material. You can also use nuts on the other end to act as a depth gauge for flushing up the end of the bolt inside of the plug hole. **[KH]**

Keith Harker

⭐ CHAMFERING TO CREATE MORE WELDING SURFACE

YouTuber Bob Clagett shares several good welding tips in a video build of a metal and wood table for his backyard. Where he attaches the square metal tubes of the table's legs to the metal frame, he chamfers where the leg is going to be butt-welded to the frame. (In case you don't know, a chamfer is a transitional angled edge between two adjoining right-angled faces of an object). The chamfer creates a recess, an extra bit of weldable surface to

ensure a solid, sturdier weld. He also points out that you should spot-weld around your entire workpiece first and then go back and completely bead the weld. This can help prevent warping of the metal as the surfaces are unevenly heated. **[BC]**

⭐ BUILDING A SIMPLE SHOP FORGE

If you've ever thought of getting into knife forging or any type of blacksmithing, a great way to get started is building your own small forge. There are a ton of YouTube videos on building forges. A popular, super low-cost method is to make one out of a paint can (where you make a furnace inside the can with plaster of Paris and sand cement walls). Of all the simple forge projects I've seen, my favorite comes from Scott Wadsworth at *Essential Craftsman*. Scott built a very well-designed and serviceable forge using little more than lightweight fire brick, some angle iron, bar stock, threaded rod, and other hardware. He finished it off with a burner he bought online for about $60. Basically, all you're doing is building a box of fire brick held together with angle iron and threaded rod. You can build such a forge with basic tools like a hacksaw and a drill. You can follow Scott building his forge here: youtu.be/TS7wumQtOs8. **[SW2]**

★ USING BOLTS AS A THREAD MASK

You can also prevent welding spatter from blocking or seizing up nuts by temporarily threading on a couple of nuts as a mask to protect the threads as you weld. Just don't weld the nuts together! **[DoRite Fabrication]**

★ FORGING YOUR OWN PARSER DRILL

Here's some fascinating proto-CNC cutting from the 19th century. The parser (or passer) drill was a bow-type drill, held against your belly, that used a template to cut a shape into wood (for things like inlay work). Neil Paskin, of the YouTube channel *Pask Makes*, tried his hand at forging his own parser drill and tested it out on several template designs (that he also made). He accomplished this with simple tools (blow torch, belt sander, anvil). The resulting drill bit, which can be chucked into any electric drill, works beautifully. Check out Neil's video to see the drill's profile and how he forged his bit: youtu.be/005_HNv_v4w. **[NP]**

FIGURING THINGS OUT ON MY OWN

I was at a maker event once, taking a class in woodworking. I made this little Japanese-style stool with through-tenons and needed some pins to keep the tenons in place. I love mixing materials, so I walked over to the blacksmithing area where a friend of mine was teaching basic forging skills. I told him my idea to forge some U-shaped pins and looked at him questioningly for guidance.

He handed me a piece of metal stock and a hammer and pointed me towards an anvil. That turned out to be the best thing he could have ever done. He knew I had forged once before and he had faith I could figure out how to do it, even though I didn't think so. The first bracket failed. As I showed it to him, he shrugged and said "Just try again." So I did.

The second one took some time, but eventually worked out. The third one I did in two heats, which is pretty much the bare minimum for what I was doing. I was so happy and proud of what I'd achieved, and showed it off with a big smile. Then it hit me that I would have never felt this way if he hadn't had faith in me. He allowed me to figure it out on my own, which was so much more gratifying.

—**Ellen Meijer**

Adobe Stock - ginettigino

16
RESTORATION

It is perhaps in the spirit of our anxious and uncertain time that antique tools, machinery, and toy restoration are becoming increasingly popular among makers. There is something oddly comforting and therapeutic about taking the old, the forgotten, the previously reliable (now seized with time, rust, and neglect) and lovingly bringing it back to life.

It is no wonder that restoration videos are all the rage on YouTube these days. The videos are usually simple, quiet (no spoken narrative), and most of the restoration process is meticulously shown, from disassembly to cleaning, sanding, and repainting to re-assembly and testing. This is a world in which a little investment in time, some Evapo-Rust, a wire wheel, and a rattle-can of enamel paint can bring the past back to its near-showroom luster.

What follows are some of my favorite restoration tips that I have recently encountered.

⭐ ACCESSING ONLINE TOOL MANUALS

You may not be aware of the fact that you can access an impressive array of service and repair manuals. Simply do a web search on the make and model of your device or tool. I rarely search out a tool or machine that I don't find. Some are on the websites of the manufacturers, others are in archives (mainly of antique tools). Sites like vintagemachinery.org are good places to look for old devices. They often have schematics, manuals, old tool catalogs, and more. These resources can be invaluable when trying to rebuild something, source parts, or just understand how something worked.

⭐ SOAKING OUT RUST

On Jimmy DiResta's Instagram feed, he shared a story about the fate of his brand new SawStop table saw. He had it under a tarp in his shop for a long time. Water leaked through a hole in the tarp and settled on the table. When he finally uncovered the saw, the surface was badly rusted.

To get the rust out, he was told to soak rags in rust remover. He only had paper towels, so he used those and some Evapo-Rust. After giving the surface a good soak and a light sanding, he got most of the rust out and the table back (minus some pitting he'll just have to live with). **[JD]**

⭐ CHEAP RUST-REMOVAL FORMULATIONS

Evapo-Rust might be the go-to brand of commercial rust remover among makers, but there are home alternatives. Evapo-Rust is a chelation agent, not an acid or a chemical. It is a synthetic iron molecule suspended in water. The chelation process forms a bond with the rust and suspends it in solution. Evapo-Rust is environmentally safe and is extremely effective. But somewhat cheaper alternatives include muriatic acid, citric acid, and white vinegar. BTW: Regardless of what testimonials you may have heard, Coke and Pepsi are not good for rust removal.

LANOLIN... FRESH FROM THE SHEEP!

Rex Burkheimer sent me this tip when I asked readers of my newsletter, "Gareth's Tips, Tools, and Shop Tales," for advice on rust prevention:

My hobby is restoring old machine tools — lathes, milling machines, drill presses, metal shapers, etc. I'm a big disciple of Evapo-Rust for rescuing old iron, but how to keep my finished projects from returning whence they came?

My shop is an unheated metal building in humid North Texas, a perfect breeding ground for iron oxide. After opening the shop to discover a big ugly rust scab on my virgin mill table, I decided to get serious.

After many trials and disappointments, I settled on... lanolin... fresh from the sheep! The story goes that Australian sheep farmers noticed that steel things that sheep often rubbed against did not rust. Sheep's wool is thick with lanolin, which is removed in processing the wool. Soap makers use it, and it's used in cosmetics.

One pound of anhydrous lanolin from eBay costs about $16. I mix it 10:1 with light oil and apply it with a brush or cloth. I also mix it with WD-40 in a trigger spray. A pound goes a long way. Since I started using this mixture about 10 years ago, rust is a thing of the past (except where I failed to apply the mixture). So far, it has been 100% effective.

—Rex Burkheimer

⭐ ALTERNATIVES TO WD-40

WD-40 is a penetration oil found in most workshops. It does a fine job, but there are some downsides. It's not nearly as cheap as home-brewed formulations and it's not as environmentally friendly. One homemade alternative is a mixture of 4-parts charcoal lighter fluid, 4-parts mineral spirits, and 1-part lightweight motor oil. Keep it in an airtight container so it won't evaporate. Another homemade recipe is a 50/50 mix of acetone and transmission fluid.

⭐ LEAD PAINT TESTING KITS

Using a lead testing kit is a smart thing to do if you're going to be restoring any old tools. You can get test swab kits relatively inexpensively online. You swab the paint and it turns a certain color to alert you to the presence of lead.

⭐ TAKE PICTURES OF YOUR DISASSEMBLY PROCESS

When taking apart a tool or machine for cleaning and restoration, don't forget to use your phone to record the disassembly process. That way, you'll have the photos to refer back to if you forget how everything goes back together.

⭐ USING SOCKETS TO REMOVE BALL BEARINGS

Removing old bearings, shafts, and pins can be tough if the penetrating oil and rust remover refuse to do their jobs. One thing you can use to hammer out these stubborn parts is a wrench socket sized to the diameter of what you're removing. You can place a rag over the socket or use a rubber- or plastic-tipped mallet to protect the socket from damage.

⭐ FREEING FROZEN BOLTS AND SCREWS

If you have seriously frozen bolts or screws that you can't mechanically remove, a quick and careful blast with a blow torch will often give them a thaw.

⭐ MAKING YOUR OWN JAPANNING

Japanning is black tool finish commonly found on hand planes, antique sewing machines, and other older tools. It is a European interpretation of traditional Japanese black lacquer that dates back to 2000 B.C. There are many different recipes for making your own Japanning. On *Hand Tool Rescue* (youtu.be/ SBqgpdBNrt8), Eric tested a number of formulae and ended up recommending a mixture of 50% turpentine, 30% asphaltum/ gilsonite, and 20% boiled linseed oil.

⭐ WHEN TO OIL, WHEN TO GREASE?

When putting a tool back together that requires lubricant, the question is often what to oil and what to grease. The rule of thumb is to grease hard-to-reach and hard-to-maintain places because grease will stay in place longer. Use oil on areas where regular maintenance is not a problem. Grease has a tendency to collect swarf (chips and filings), so oil is better in situations where a lot of material is going to collect. **[EK3]**

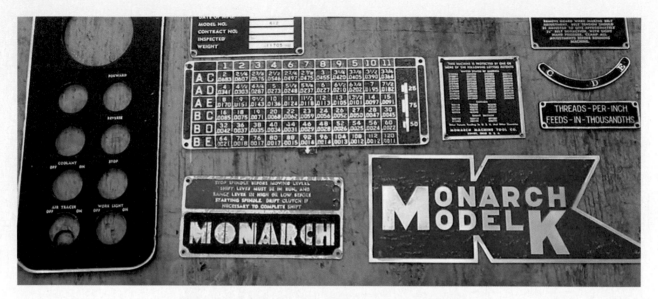

⭐ CREATING LABELS AND BADGES FOR RESTORED TOOLS

Once you've repainted an old tool, it really finishes the restoration to re-apply the original labels and badges (or at least reproductions of them). There are many antique tool and machinery labels available online, so look before you try and recreate your own in a vector drawing program. Check places like vintagemachinery.org's Decals section. To give you some idea of how this process of restoring machinery badges works, check out the video *Vintage Machinery* did: youtu.be/AxQJT0aSYVI.

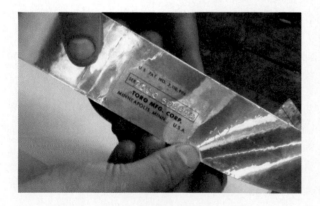

⭐ RECREATING SERIAL NUMBER PLATES

Eric of *Hand Tool Rescue* has a clever way of recreating serial number plates. He first punches the original serial number into metal foil tape using a number punch set. Then, he prints a replica of the original label onto clear adhesive label paper and affixes that over the foil. Cut out and apply. Genius!

GIVE THE GIFT OF RESTORATION

I have recently gotten into some restoration work myself, mainly old hand tools. I have given several to my wife as Christmas presents. One was a very beat up old hammer that she had in her art studio. She'd needed a hammer one day, couldn't find her normal one, so she walked down the street to a Goodwill and bought one there for 99 cents. It looked like it had been to war and back. It was covered in paint, spackle, and adhesive, and the handle was split near the head. I "borrowed" it from her studio and spent some nights after dinner bringing it back from the dead. I cleaned and sanded the handle, epoxied the split, and sanded and buffed the head. A coat of black enamel on the non-business surfaces of the head and some Walrus Oil (not made of walruses) on the handle, and the hammer was almost as nice as the day it was made. I gave it to her in a box that included the before photo. She loves it.

If you have a loved one that uses tools a lot, consider refurbing their tools (with their permission, of course) or buy, refurb, and give antique tools as gifts. Many older tools are much better made than tools of today and they're a lot cheaper. Yard and estate sales are great places to pick up old tools for a song.

⭐ CLEANING OLD HARDWARE CASES WITH "RETROBRIGHT"

Retrobright (aka Retrobrite or Retr0bright) is a DIY hydrogen peroxide formula that can be used on old computer, game console, and other consumer electronics cases to bring back some of their original color.

On his website, John Edgar Park shows off an old 1989 Nintendo GameBoy case that he brought back to life by using hydrogen peroxide at 6% concentration, Oxi laundry booster, and UV light. As he points out, instead of these separate chemicals, all you need to do is get some salon-grade cream hair bleach. Either way, you cover your case in the chemical mixture, wrap it in plastic wrap, and leave it outside in the sun (or under a UV lamp) for six hours. The results are quite impressive. **[JEP]**

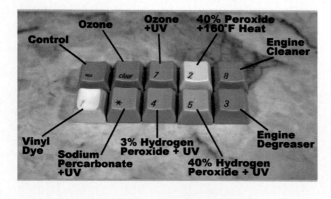

⭐ MORE ON USING RETROBRIGHT

Dave Murray, aka *The 8-Bit Guy* on YouTube, has experimented with a number of different formulations for Retrobright (beyond the common hydrogen peroxide, laundry booster, and UV light). The surprises in his testing were ozone and UV light and hydrogen peroxide and applying heat. For Retrobrighting large objects, pumping ozone into a clear container and subjecting it to sunlight looks like a sensible approach. For more details, see Dave's YouTube channel: youtu.be/qZYbchvSUDY. **[DM2]**

★ FIXING CORRODED BATTERY TERMINALS

Who hasn't opened an old piece of consumer electronics that has batteries in it that have long ago corroded out? If you've ever wondered what to do to properly clean out the battery bay and restore the terminals, all you really need is some vinegar to clean away the potassium hydroxide that has leaked from the batteries and a wire brush or light sandpaper to clean the contacts. If things are really bad, you can desolder the wires and remove the terminals for better cleaning, or you can replace them if need be. Where do you get new terminals? From old battery-powered devices you are no longer using. For more detailed instructions, see lonesoulsurfer's Instructable: instructables.com/ How-to-Fix-Corroded-Battery-Terminals.

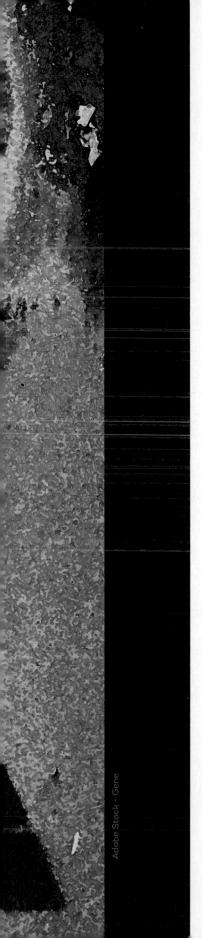

HOBBY TIPS

Of course, many of the tips in this book are "hobby tips." Regardless of what you're making, it likely involves some form of cutting, fastening, gluing, finishing, etc. That said, there are techniques that are specific to various forms of recreational making and crafting.

What follows are some of my favorite hobby-specific tips. Many of them first appeared in "Gareth's Table of Tinkering," a column I wrote a few years ago in *Bexim's Bazaar*, a "game crafting" magazine. But these techniques can also apply to diorama building, cosplay (costume play), scale modeling, Halloween projects, prototype building, faux finishing, and much more.

★ LIQUID STYRENE AND STYRENE CARVING

Make: contributor and toy designer Bob Knetzger sent me a number of tips on working with styrene. He writes:

"Save your clean styrene shavings. Add some MEK (methyl ethyl ketone) to them in a small glass bottle to make a styrene slurry. Cap it very tightly when you're done and save for later. You can use 'liquid styrene' to build up radii, fill inside corners, or mend gaps. It sometimes works better than the super glue and baking soda trick.

You can build up thicker, 3D shapes with layers of styrene sheet, then scrape and Dremel them to your final shape. If you're only making one model, it's faster than sculpting and casting. **[BK]**

Bob Knetzger

⭐ EVEN MORE STYRENE HACKS

Bob Knetzger also shares this tip: "Don't forget about the thermoplastic properties of styrene. I made this fun *Bug Boys* PEZ for my daughter, Laura, on her birthday. (*Bug Boys* is the name of her comic book and this is one of the book's characters.) I took an old Garfield PEZ dispenser, cut the ears off, then reshaped the round head a little on a disk sander. Because PEZ dispensers are made of styrene, I could make up the added horn parts out of laminated sheets of styrene and just solvent-bond them onto the PEZ head (much stronger and more durable than super glue). Before I bonded them, I scraped the needed radii, then used a heat gun to soften the main horn so that I could bend it into the final curved shape. Came out great!" **[BK]**

⭐ HOBBY VISE

Last year, I broke down and bought a Dremel Multi-Vise on Amazon ($35). I have always wanted one, but thought it was a somewhat frivolous purchase. I absolutely love it now and have found it surprisingly versatile. The wide, rubber-padded adjustable jaw (13 cm), the grooves on the pads for accepting round stock, the ball swivel base, the adapter to hold a Dremel tool. So many useful features. I even use the bar that the jaws travel on to steady my hand as I paint. Now I wish I hadn't gone so long without one.

⭐ USING PAINT AS GLUE

For small, lightweight parts, you can actually use paint to "glue" the pieces onto a "painting stick." Many modelers use home store paint stirring sticks to hold their minis in place for priming (you can also use a strip of stiff cardboard). To hold the items in place, you just lay down a quick coat of primer on the paint stirring stick, press the parts to it, and your parts are secure enough for priming (and for easy removal). For heavier items, a tiny dab of hot glue (or CA glue) is more secure. When the painting is done, use a razor knife to gently pry up the parts.

⭐ PAINTING OUTSIDE THE LINES

One thing that I think novice painters are often too reluctant to do is paint "outside the lines" when painting a small model or figure. They will tensely and frustratingly try to paint everything perfectly from the first coat, and become frustrated when they inevitably get paint where it's not supposed to go. Because you are going to be painting multiple coats, and likely using undertone washes, you will have plenty of time to fix any boo-boos. Relax, have fun, and make sure to fully cover things like the edges of belts and straps and other raised details. You can clean up the edges as you work on neighboring colors.

⭐ PAINT CRACKLE

On the YouTube channel *Grimdark Compendium*, in a video on painting a 40K Mechanicus Dune Crawler, Ryan Oates demonstrates an interesting way of getting a crackling effect on paint to make it look aged and decayed. You first apply a coat of what's called chipping medium (he uses AK Interactive's Heavy Chipping Medium) over primer and a layer of satin varnish, and then you apply your base coat over that while the chipping medium is still wet. When it dries, it will create cracks in your paint. youtu.be/ FCvLA93yWNE **[RO2]**

⭐ GLOW EFFECTS WITHOUT AN AIRBRUSH

On his *Midwinter Minis* YouTube channel, Guy Perchard shows how to get a very nice glow effect with little more than a drybrush. Basically, what you do is paint the area you wish to have glow with a dark shade of your base color (in Guy's righthand example, blue). Then, you drybrush the area and beyond it (places where the glow of light would hit) with progressively lighter shades of your base color, ending in a light dusting of white. Easy and effective. youtu.be/qkLZZbndav8 **[GP]**

WORKING WITH STYRENE PLASTIC

Anyone who's built a scale model kit has worked with polystyrene plastic (often shortened to "styrene" or abbreviated as "PS"). This ubiquitous thermoplastic is molded into everything from disposable product packaging to CD/DVD "jewel boxes" to model kits and toys. But what many makers may be aware of is the speed and versatility of this material for mocking up three-dimensional designs — for prototyping, creating custom project boxes, and for scratch-building models. It can also be used in vacuum-forming. With a few simple tools, a pot of cement, and a bit of practice, you can render a design in 3D with impressive speed. If properly engineered, the resulting object can be surprisingly durable and strong.

ACQUIRING STYRENE

Polystyrene stock can be purchased in a dizzying array of sizes, thicknesses, shapes, and textures. You can buy sheet, tube, rod, strip, and architectural-detail stock directly from well-known manufacturers like Plastruct (plastruct.com) and Evergreen Scale Models (evergreenscalemodels.com), at online retailers, or from your local hobby or game store.

If you're new to buying styrene stock, you might want to see what you're getting physically to choose the types of thicknesses and shapes you're interested in. To build up a supply, Evergreen sells assortment bundles of sheet stock, tubes, and rods. Common sheet sizes are 6"×12" and thicknesses of .01" (.25mm), .02" (.5mm), and .04" (1mm). Rods, tubes, I-beam, T-beam, L-beam, and other shapes come in various sizes at 14" length.

If you start using styrene a lot as a prototyping/building material, besides acquiring a supply of new stock, you can also start collecting polystyrene-based packaging, sprue material from model kits (sprue is the plastic frame that holds small extrusion-molded parts), and model kit parts to incorporate into your creations.

Once you gather the simple tools you need and become proficient at marking, cutting, gluing, and sanding styrene, this old-school technology can be a worthy competitor to 3D printing for creating quick prototypes.

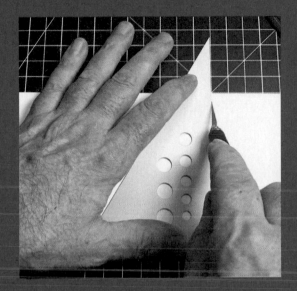

SCORING AND BREAKING

One of the great benefits of styrene is that it is a score-and-break plastic. You don't have to cut through it all the way with a razor knife, you simply have to score it a few times, and then you can gently bend and break it cleanly along the score line. For thicker stock, it is recommended that you use a plastic cutter (or linoleum knife) for deeper scoring. Square-nose pliers are also helpful for creating a clean break along the score line of thicker stock. Once you have a design plan for what you are constructing, and have precisely marked out your pieces, you can quickly and accurately score and break all of the plastic components you need.

DEBURRING

After scoring and breaking your pieces, especially with thicker stock, you'll want to check to make sure that you have clean cuts, without burrs. If you do find excess plastic along your breaks, you can use perpendicular strokes of a hobby knife blade (or a deburring tool) to clean everything up before gluing. A light sanding of the edges can also clean up any excess.

BASIC CONSTRUCTION

Polystyrene is an impressively strong material for its weight, and it can be very flexible in smaller thicknesses. To maximize its structural integrity, you want to think about reinforcing your design as you would when prototyping/modeling in cardboard, wood, other plastic, or any other material. Styrene is very forgiving to work with. As professional model-maker (and MythBuster) Adam Savage likes to say, "styrene hides a ton of crimes." It is a material that's perfect for precision construction, but it is also great for improvisational designing and scratch-build model-making. For a perfect example of this, check out Adam's YouTube build (youtu.be/ZfvtGrhYk0I) where he scratch-builds an interpretation of a Mobius sci-fi illustration in a day. This video is a styrene building masterclass, with tons of great ideas, like using the tip of a sharp hobby knife blade to pick and place small styrene parts onto your model.

BEND ME, SHAPE ME

Because polystyrene is a thermoplastic, you can easily heat-bend it to achieve many desired shapes. In hobby modeling, a common technique is called "sprue-bending." This is where you take a piece of leftover model sprue, hold it over a candle flame, and bend it to get the desired shape you want. With a little practice, you'll be amazed at what sorts of parts you can fabricate with scrap or stock styrene rod.

GLUING STYRENE

There are a number of glues and solvents you can use with polystyrene. The most common is plastic model cement, which is a mixture of polystyrene dissolved in toluene. The drawback here is a longer set time. Using cyanoacrylate (CA) glue can greatly shorten the wait and allow you to build/prototype much faster. What professional model-makers use is a solvent cement, such as Weld-On. This can quickly be brushed on or applied with a needle applicator. Professional solvent cement is rated by drying times. Weld-On #3 has a one-minute workingtime, a two-minute fixture time, and is a common choice for quick constructions. The drawback to solvent cement is that it is extremely flammable and requires good ventilation. For structural integrity, it beats all glues because it literally fuses the plastic pieces together.

BUILDING A SOLVENT WELL

In the movie biz, where solvent cement is the go-to adhesive for model-making, fashioning a solvent dispenser is de rigueur. To create one, all you need to do is cut off the bottom of a soda can and glue it to a thick piece of styrene card. Then, you simply pour a working supply of cement into the well and use a glue brush or needle applicator to apply the solvent to your plastic joint.

SANDING

One of the other great benefits to working with polystyrene is that it is easily sanded. If you "work proud and sand down" ("proud" meaning to let the pieces protrude a little), you can quickly create smooth and perfect edges, hide seams, etc. A good complement of sandpaper, sanding block, and sanding sticks, at varying grits, will allow you to quickly remove material and work your way up to a desired smooth finish.

FINISHING

To get the best paint results, it is a good idea to clean your polystyrene constructions with alcohol (or just a light bath in mild detergent water) before painting. Polystyrene can be primed with spray (or brush-on) acrylic, enamel, or lacquer primers, and finishes beautifully with brush-on or air-brushed acrylics. The composition and finish of styrene is such that you can achieve a wide range of faux-finish and weathering effects with this material.

Gianfranco GavazziFisher

⭐ MODELING WITH TRASH

My latest modeling obsession is a Facebook group called "Trash Bash Bits." Members show off amazing models that they've created (and how they got there) using little more than kitchen trash. You'll never look at plastic bottles, food containers, coffee stirrers, deodorant tubes, and toothpicks the same way. To the left is is a work-in-progress image by group member Gianfranco GavazziFisher. It is made of miscellaneous trash pieces, electronic parts (including a dead camera), and a styrene ball. **[GG]**

⭐ DOES YOUR CREATION TELL A STORY?

If you're making a model, diorama, cosplay costume, or even doing things like faux finishing or weathering something, don't forget that wear and weathering on an object should tell a story. How did the object get used? What elements was it subjected to? Where would it show wear? Signs of abuse and repair? Think through the life of your object, and wear it down and weather it accordingly.

⭐ PASTEL ART CHALKS FOR WEATHERING

Hobby supply manufacturers sell all sorts of bottled weathering chalks that you can use to add dirt, rust, mud, and grime effects to your models. These products can be quite expensive. This is especially the case when you realize that all that's in those overpriced bottles is pigment-saturated chalk. You can achieve the same results by buying a set of cheap art chalks and using a knife or mortar and pestle to turn the chalk stick into a powder. From there, you can apply the chalk to your model with a soft make-up brush. One important note: Chalk won't stick to gloss finishes very well. You'll need to hit the area with some dull coat first. And, of course, when you're finished, you'll want to seal and protect the weathering (and the rest of your model) with another layer of varnish. You can get big sets of art chalks (48 colors) online for under $10.

⭐ ART PENCILS FOR WEATHERING

Along with weathering chalks, hobby suppliers have started selling weather pencils. Like the chalks, these are little more than regular art pencils in the colors of rust, grime, dirt, etc. You can get the same results using conventional art pencils in the colors you desire. The above photo shows one of my gaming miniatures that I finished using art pencils in white and shades of brown to do edge highlighting. I used art pencils I already had in my supplies, but you can get a set of them online, 48 for under $20.

⭐ MAKING WAVES WITH AIR

Roman Khramov, a YouTube diorama maker, shows how you can use an empty airbrush to make waves in a commercial water effects medium, which dries clear. This also works using the more common Mod Podge. Put down the medium you're using and blow it with an empty airbrush to create a wave effect. If you've never thought of using an airbrush as an air tool like this, now you will! **[RK2]**

⭐ FREEZER PAPER IMAGE TRANSFER

To create a printed image on wood (you can experiment with other materials), print a mirror image of your art onto kitchen freezer paper (cut to fit into your printer), and then burnish it (rub it down) onto your workpiece before the ink on the paper dries. I'm anxious to try out this technique. Spotted inthe dice tower project on the *Wicked Makers* YouTube channel.

⭐ TISSUE PAPER PRINTING

I'm in the process of building some 20mm billboards for a tabletop gaming board. In doing research on how best to create the billboard art, I ran across a method for printing art onto tissue paper (that's been secured to a regular piece of paper and run through your printer). Remember to print with the shiny side against the carrier paper. Following this method and then securing the tissue to a piece of balsa wood with spray adhesive gives your billboard a realistic, painted-on, and faded effect.

⭐ SPIDER WEBS FROM DRYER SHEETS

On *Black Magic Craft*, in a video on "Spider Web and Egg Terrain", Jeremy Pillipow shows you how easy it is to make realistic-looking cobwebs out of dryer sheets. The process involves little more than tearing apart the used dryer sheet and gluing it down to your base or terrain piece to create a suitably creepy blanket of spider web. **[JP]**

⭐ USING WHITE GLUE AND TISSUE PAPER TO MAKE TARPS AND ROLLS

If you've watched any *DM Scotty* videos on YouTube, you know that he loves using the brilliant technique of combining toilet paper/tissue/paper towel and watered-down PVA/white glue to create all sorts of moldable forms for game crafting. Back in the aughts, I used this method to create bed and tent rolls for some of my sci-fi vehicles. To create a tarp, soak tissue paper with a 50/50 mix of white glue and water and drape over the vehicle or whatever you wish to cover. After it is thoroughly dry, apply paint, and *voilà*. To create tent and bed rolls, simply cut the size roll you wish in tissue paper, soak in a water-glue mix, and then carefully roll it up. You can use thin wire or thread to create ties. Properly painted, the results are surprisingly realistic. **[DMS]**

⭐ USING PLIERS TO MAKE WOOD GRAIN

On the amazing YouTube channel *Real Terrain Hobbies*, in the "Realistic Ocean Diorama" video, Neil shows how to use a set of pliers to pinch faux wood grain into wooden dowels. The results look quite effective. Note that he angles the jaws back and forth to get a wavy grain effect. If you haven't checked out Neil's channel, you are in for a treat. He makes the most jaw-dropping pieces, including a medieval/fantasy tavern where he has laid every (foam) stone and (balsa) timber by hand, and his current project, a wizard's tower rising from a stormy ocean.

⭐ MAKE YOUR OWN RUST WASH

I just discovered this tip recently and it's a keeper. To create very realistic rust on your cars and terrain, use... rust! To make it, all you have to do is soak some steel wool in white vinegar for a few weeks (or longer). The wool will disintegrate into the vinegar and turn a lovely rust brown color. I put some steel wool and vinegar into an old paint pot and let it rot for a month (shaking occasionally). To apply, brush on with a trash-brush. One note of caution: The wash goes on looking like water (the rust "pigment" is not very visible). When it dries, the rust will remain where the wash has pooled. This makes it a very realistic looking weathering effect. But it also makes it hard to control and easy to overdo. So, go slow, and maybe do a test-wash on something first.

⭐ CINNAMON RUST

By way of Adam Savage's *Tested* comes another offbeat idea for rust effects: cinnamon. In a Tested.com video with Norm Chan and Kayte Sabicer, called "Weathering a Model (Space) Ship," model-maker Kayte shows how to use regular kitchen cinnamon, applied to acrylic paint, to create a rust effect. She recommends painting a rust-colored acrylic paint first and then, while the paint is wet, using a big brush to flick the cinnamon onto the painted area. As she points out, rust isn't a color, it's a material process; it has substance. Cinnamon

can add some of that substance to create a more realistic rust look. I get a similar effect by using the powder from reddish-brown art chalk. **[KS]**

18

TROUBLESHOOTING & MAINTENANCE

When faced with a repair problem — any sort of problem, really — it's important to have the right troubleshooting mindset. First off, always keep a positive attitude. You can figure this out and fix it. Be methodical. Break the problem down into smaller parts. Try to not make too many assumptions. Start with the simplest, most likely reasons and work it out from there. If the process starts to get too frustrating, take a break and come back. Don't let the problem wear down your resolve. And, if you truly get stuck, bring in help. A fresh set of eyeballs and assumptions can often identify something that you missed. Be curious about how the machinery in your life works, know how it functions, and set aside time each day/week/month to maintain it.

★ CONSULT THE SERVICE MANUAL

Manuals to almost anything can be found online. It can often be very helpful to look at a service manual if you're encountering a problem with a tool or machine. Sometimes, the very problem you're having is addressed in a troubleshooting section. Or, just looking at the technical drawings of the machine can help you take it apart, put it back together, and to better understand what's going on. Also, many manuals have a maintenance section which details how often to replace parts and lubricate the machine.

⭐ BREAKING THE PROBLEM DOWN

See the broken device as a series of subsystems; isolate and check each one to rule it out as the source of the problem. Ruling out what isn't wrong can help you zero in on what is. Run through the entire operation of the device as you move through the actual machine. Test parts as appropriate (using a multimeter, mechanically, etc.).

⭐ WAVING A DEAD CHICKEN

Sometimes, when faced with a particularly vexing problem, you start doing irrational things after seemingly exhausting all of the rational ones. In hacker jargon, this high-tech voodoo is referred to as "waving a dead chicken." The crazy thing is, sometimes, woo-woo dead chickens resolve actual problems. Recently, I had a device that was not being recognized on a network. I did the obvious "troubleshoot" of turning it off and back on. I waited. Nothing. Several minutes later, I did it again, just for good measure. Again, nothing. Now, logic would tell you that if it didn't show up after two power cycles, there was no need for a third. But I 'd tried everything else I could think of and was ready to call the device faulty. But I cycled power one more time and lo and behold, it blinked into existence on the network. And it's been trouble-free ever since.

⭐ CLEAN AND SERVICE YOUR TOOLS OFTEN

It would be ideal if we cleaned and maintained our tools every time we used them. For most hand and power tools, such cleaning and maintenance only take a few minutes, but most of us mere mortals have a hard time disciplining ourselves to do this. We want to finish the job, move on to something else, anything other than getting out a rag, a vacuum, an oil can, and a sharpening file. If you can't get into an every-use cleaning habit, set a regular calendar alert or set aside a day each week (or month) where you do tool cleaning/servicing. For decades, I've used Sunday as my "system reboot" day. It's the day when I plan for the coming week, order supplies, and clean and organize my workshop. This is the perfect time to also clean and organize tools.

Adobe Stock - Gresei

★ STAY SHARP

Keeping your tools sharp not only helps them do their jobs better, it also extends their life. Sharp cutting tools are safer to use because they allow for greater, more predictable control and don't require so much pressure. Tool sharpening is another task that many makers regard as some sort of dreaded dark art. It isn't. Most of what you need (e.g. a vise, files, oil), you likely already have, and special sharpening guides and stones aren't that expensive, either. As usual, YouTube to the rescue with excellent tutorials on sharpening everything from saw blades and mowers to scissors and shovels. And whatever time and money you spend sharpening, you can remind yourself that you're likely making that back in extending the life of the tools and equipment you've invested so much money in.

★ COOL IT!

Besides sharpening tools, another way to extend their lives is to let their motors cool down as they get hot. This might seem obvious, but many users of consumer-grade tools seem to overlook this preventative measure. Use basic common sense. If the motor is getting especially hot, give it a rest. If you're cutting through dense, hard material like concrete, give the tool a rest every few minutes as you go. A cool, well-lubricated tool is a happy tool.

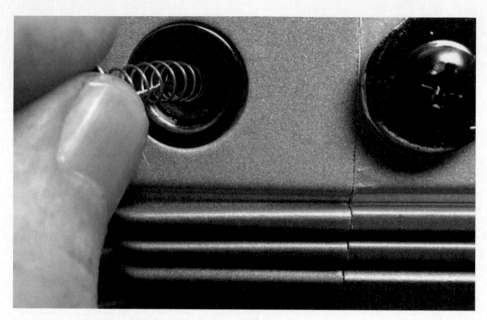

⭐ FIXING YOUR ROTARY TOOL

Sadly, many rotary tool (aka Dremel) owners don't seem to know that when a tool becomes sluggish or stops working, it is likely only due to worn-out motor brushes. These are easily and cheaply replaced via two screws usually found on the sides of the tool. The carbon brushes can be bought online for under $10 and they literally take seconds to replace.

⭐ POURING OIL WITH A SCREWDRIVER

Here's a little trick if you find yourself without a funnel and need to pour something, like engine oil. Slowly pour down the side of a screwdriver. The shaft of the driver will act as a guide for the oil. It's funky, but in a pinch, it'll get the job done.

Adobe Stock - siam4510

19
SAFETY AND FIRST AID

Safety and first aid might not be the most exciting aspects of making, but you ignore them at your peril. Often, the more proficient you become at a trade, the cockier and more careless you can behave. Jack Parsons, the self-taught chemist and rocket scientist who co-founded the Jet Propulsion Laboratory, became so convinced of his mastery over explosives that he relaxed his guard (and his grip on a beaker of mercury fulminate) and died in a violent home lab explosion.

If you work with anything that cuts, spins, splashes, shocks, burns, or combusts, or if you engage in activities, places, or processes that could harm you or others — in other words, if you're a maker! — you need to always remain mindful of the risks. And the more you know about safety precautions and emergency first aid, the more you can confidently engage in your work with the knowledge that you know what to do should something go wrong.

Here are a few safety and first aid tips that have come across my desktop since the publication of the first *Tips and Tales from the Workshop* collection.

⭐ FEELING FAINT? LIE DOWN!

If you do feel faint, lie down. It's far less embarrassing (and dangerous) to lie down and recover than it is to fall down and lose control of your bowels. **[CL]**

⭐ BASIC SAFETY GUIDELINES IN THE SHOP

Here are the basics for conducting yourself in the workshop:

- Always use eye protection if you're doing anything where material can fly, splash, or aerosolize.
- Always use ear protection if you're doing anything that generates loud noise.
- Always wear a respirator if you're doing anything that generates airborne particles or fumes.
- Always wear closed-toe shoes, ideally steel-toe shoes or boots if you're working with anything heavy or sharp.
- Always secure hair and avoid (or secure) loose-fitting clothing.
- Always read the safety instructions and warnings on materials and tools.
- Clean up as you go as a safety practice.
- If you're working in a space with others, clean and disinfect your tools and work area when you're done.

⭐ HANGING SAFETY GEAR BY THE FRONT DOOR

Engineer, maker, and YouTuber Jeremy Fielding recently moved and had to set up his new workshop. One great safety and shop organization idea that he shared: hanging all of your ear and eye protection by the shop door so visitors to the shop can grab them as they enter and return them to the hooks as they leave. **[JF]**

⭐ YOUR TOOLS ARE TRYING TO MURDER YOU!

Quinn Dunki, aka BlondiHacks, reminds us: Your tools are trying to murder you! Okay, maybe that's a tad overly dramatic, but the point should be well-taken. You don't really want to fear your tools, but you want to be mindful of the dangers they pose and know that they are ready to harm you if you don't maintain a healthy respect for those dangers. **[QD]**

THAT TIME I ALMOST BECAME A TRANSFORMER

As far as safety goes, I suppose there's the reason why I work with high voltages one-handed. I was working on an old Super 8 projector (I wanted to watch my copy of *Alien*), and I switched it off while I was messing around inside.

Unfortunately, being made in the 1960s, safety was not job #1. The switch cut the power to the low voltage side of the transformer, but not the high. I had my left hand on the metal case, and my right hand on the transformer's high side. Next thing I knew, I was flat on my back on the floor. 240 volts straight across the chest.

Now, I keep one hand in my pocket so that I never give electricity a direct path across my chest, and I never trust a switch.

—Andrew Lewis

⭐ WHY THE END OF THE DAY CAN BE DANGEROUS

The last job of the day is the most dangerous. Too many people, after 5 p.m. on a Friday night, doing "one last thing" before going home, get themselves into trouble. That's the time when concentration wavers and you screw your hand to a plank, fall off the roof, or really wish you'd brought that SawStop table saw. Never try to finish a job in a rush or at the end of the day if you're feeling tired. **[CL]**

⭐ VENTILATION IS IMPORTANT, BUT SO IS LIMITED SKIN CONTACT

Everyone knows that you should use solvents in well-ventilated areas, but respiration isn't the only way solvents can get into your system. Prolonged skin contact can also be a problem. Wear gloves and a long-sleeve shirt and long pants. The less of you exposed to strong solvents, the better. **[AL]**

★ STAY WELL-VENTILATED AFTER THE JOB (OR LEAVE YOUR CLOTHES BEHIND)

Working in a well-ventilated area is essential when working with solvents, but staying in that well-ventilated area for a while after you're done is important, too. The clothes you're wearing can soak up the vapour and release them when you move to a confined space. This is also true of other chemicals, like chlorine, which can seep into your clothes and continue to outgas for some time afterwards. Let the space and your clothes air out before leaving it, or leave your work clothes in the shop if you can. **[AL]**

★ SHOP FIRST AID HAS TO BE ACCESSIBLE TO THE BLEEDING

Adam Savage of *MythBusters* fame, who now runs the busy website and YouTube channel *Tested*, is known for his organizational wisdom. One great tip that he's shared about first aid is to always have your first-aid supplies in a close-by, easy-to-access cabinet or bin, to know where everything is, and to have everything accessible one-handed, in case you're bleeding from the other hand. I discovered the need for this the hard way. I cut my hand badly and had to access my first-aid kit while my left hand was bleeding all over everything. I have a really nice first-aid kit, but it's stored in the bathroom closet and is in a zippered case. Trying to unzip it and get what I needed while mildly panicking and dripping blood everywhere was a real eye-opener. A much more sensible set-up I plan to put together is to have a plastic bin on a shelf in my shop that I can pull down one-handed. Adam suggests having iodine, hydrogen peroxide, cloth bandages, gauze pads, closures, and triple antibiotic cream. I would add tweezers, scissors, medical tape, aspirin, disposable non-latex gloves, and a decent emergency first-aid guide. **[AS3]**

★ DO YOU FEAR IT'S SCARY-BAD? LOOK AT IT GRADUALLY

If you're presented with a really gory injury that you're worried will make you feel faint, cover the wound as soon as possible without looking at it, then gradually expose it while looking at something else. Glance at the exposed part, reveal a bit more, glance again, and then when it's completely exposed, glance at it for a couple more seconds, then a few more, then as long as you need to. Doing this gradual reveal can help mitigate the shock of seeing a particularly nasty injury that might cause you to faint. **[CL]**

⭐ SALINE IN THE SHOP. ALWAYS!

My pal and yours, Andy Birkey, wants YOU to be safe in the shop! Especially with those precious peepers. Andy reminds us to always keep a bottle of saline solution in the shop and to know exactly where those drops are, as you may be trying to retrieve them with a mote in your eye. **[AB]**

⭐ KEEP A LIST OF ALLERGIES AND MEDICATIONS

Always keep a list in your wallet or bag of known allergies and medications that you take. You can either write or type out your list, or peel the label from an empty bottle (or medication box top), which includes the name and dosage, and stick it to a piece of paper. This will help medics in an emergency. **[CL]**

⭐ STOP THE BLEEDING, THEN DRESS THE WOUND

While it may seem counter intuitive, if you're injured and bleeding, the first thing you cover the wound with doesn't necessarily have to be a proper dressing. You're going to have to clean the wound anyway. Your first priority is to stop the bleeding. Anything reasonably clean, like a towel or cotton shirt, will do. **[CL]**

⭐ APPLYING BUG SPRAY

I was reminded of this smarter method of applying bug juice to your face via Beau Ouimette of the popular metal detecting YouTube channel *Aquachigger*. Spray the repellent on the back of your hand and then apply it to your face: cheeks, nose, chin, and forehead. This way, you avoid getting a spray in your eyes and up your nose, and you keep your hands clean(er). **[BO]**

20

MAKER VIDEO PRODUCTION

There are so many makers now doing YouTube channels, Instagram stories and reels, TikTok videos, and the like. For this volume of the series, I thought it would be a good idea to include some tips and best practices around maker video production.

I have done very little video production myself, but, as the former editorial director of *Make:*, I was the executive producer of all video content — so I developed my own set of production design ideas and standards.

And now, as a tips merchant, I look at hundreds of maker-made videos each month, so I definitely see patterns, strengths, and weaknesses of current video production approaches. Here is some of what I've learned.

⭐ AVOID HOME SHOPPING NETWORK NARRATION

This tip concerns what I call the "Send Before Midnight Tonight!" production approach. This is the kind of over-driven sales-speak that we've all heard in vintage late-night, low-budget TV commercials and QVC/Home Shopping Network programming. If you listen to something like QVC, you will hear the sales hosts starting nearly every sentence with, "You're going to want to...," "You're going to get...," etc. This is a psychological sales "trick" of making the potential buyer already imagine themselves behind the wheel of their NEEEEEWWW CAR! (Or whatever.) It's annoying enough when we hear this kind of manipulative speech in this type of manipulative sales programming, but I also hear it a lot in makers' YouTube project narration. "You're going to take your craft wire and...," "You're going to want to fold the edges...," etc. Ultimately maybe not a big deal, but something to be mindful of. Also, it's terrible English. Just say: "Take your craft wire...," "Now, fold the edges...," etc.

⭐ SWERVE INTO YOUR WEIRDNESS

This tip comes to us from YouTube star Jimmy DiResta. When giving new YouTubers advice on getting started on a maker channel, Jimmy tells them to "swerve into your weirdness." What he means is to think about what makes you unique, memorable, interesting to watch, and magnify those things. "Embrace your weirdness. Don't suppress it. Express it," he advises. This was actually advice given to Jimmy by his friend, TV actor, and host of *Making It* (NBC), Nick Offerman. **[JD]**

⭐ KNOW HOW TO "GET IN AND GET OUT"

Here's another tip for makers doing live media, public speaking, or otherwise being called upon to sound intelligent on the spot. This was shared with me by a friend who's a local DC TV news reporter. The basic idea is to know what you're going to say to open your remarks, know how you're going to finish, and then you're free to ad lib through the middle. Or, the way he put it to me was: "Know how to get in, how to get out, and then just BS your way through the middle." This is basically true in writing, too. If you have a strong concept, title, opener, and closer, most of the heavy lifting is done. Then it's just a matter of who, what, where, when, and why-ing your way through the middle.

Use code "STUFF20" for 20% off first monthly box

⭐ SMOOTH INTEGRATION OF SPONSORSHIPS

At World Maker Faire years ago, on my panel discussing YouTube makers' best practices, we talked about the different ways that makers integrate sponsorship content and ads into their videos. I pointed out how much I like the way panel-member Bob Clagett does it. While some repetitive process is happening on screen, say, repetitive cutting or sanding, midway through the video, he backgrounds that process and uses an inset window to plug his sponsor. I also like the fact that if, for example, he's plugging Audible, he talks about a book he's currently listening to on the service and recommends. It all feels seamless and non-invasive, and it motivates me to want to support his sponsor. **[BC]**

⭐ TALK ABOUT YOUR SPONSOR AS YOU WORK

Here's another unique way of doing sponsorship ads. Instead of putting it in the front, middle, or end of your video, or backgrounding a process portion and using an inset window as in the example above, try talking about your sponsor as you do that repetitive process portion, just as you might talk to a friend if you were both sitting around working on a project together. You can even work in an association with what you're doing and the sponsored project (but only if it's not a laughable stretch). I see a lot of makers doing this now when their sponsor is Skillshare. They discuss a class on Skillshare that's related to the project they're doing. You can also use insets related to the sponsor as you talk, while you work. So, in essence, you are reversing the technique that Bob Clagett is using above.

★ WORK WITH SPONSORS IN A WAY THAT WORKS FOR YOU

Sponsors are a big part of the sustainability and success of a YouTube channel. The maker space on YouTube is popular enough that, as soon as you begin to gain some audience and attention, sponsors are going to come a-calling. It doesn't cost, say, a tool company or a home store, much to offer you free tools in exchange for a mention and branding on your videos.

It is easy, however, to quickly fall into a relationship that is driven by the sponsor, and not by you. That may work for them, but ultimately, it is not going to work out well for you. Especially when you're new, you can feel like you have to do whatever the sponsor says to hold their interest in you. Before you know it, your sponsor might start making demands to you like removing all of their competitors' tools from any shots in your video, or using certain wording in how you promote their products that you may not be comfortable with. Unless you are a YouTuber with thousands or millions of followers, you may feel like you don't have a lot of leverage. You don't. But what you do have is your own integrity and sense of where you want your channel to go and what the spirit and tone of it should be. You have to be willing to speak up and be willing to even lose a sponsor if they are asking you to do things that you're uncomfortable with. The free tools, any sponsorship payments, and perks are attractive, but PR managers and firms are in the business of brand association and brand clarity. Many companies are so used to content creators doing exactly what the company reps demand that they become invasive to your mission: content creation. Don't be afraid to push back. At least a little. Create a reasonable firewall between your content and your sponsors and stay behind it! And don't be afraid to come back with an even better idea for a sponsorship campaign that serves their interests but that works best with your content (see "Having the Courage to Say No" sidebar).

HAVING THE COURAGE
TO SAY NO

Being the editorial director of a media company, you have to have a great working relationship with your sales and marketing team. I had that with sales and marketing at *Make:*. They respected me, I respected them. But sometimes my stubbornness over the established content/sales firewall would cause tension. We always worked it out and often ended up with something that made all of us feel reasonably wholesome about what we were doing. Here is a perfect example.

At one point, sales and marketing brought a deal to me that I couldn't refuse. Or could I? They'd been approached by one of the iconic bar soap companies. Said soap powerhouse wanted to do sponsored blog posts on the *Make:* site that they would write. We didn't have to do anything. Just sit back and enjoy the free content they'd provide. A bit of relief for the overworked blog team. With a decent advertising buy-in. Everybody wins, right?

No way, said I. Not a chance. We took pride in the integrity of our content, our voice, and the trust our readership placed in us. Literally selling soap on a site about making stuff seemed way out of bounds. Sales was not happy about my refusal, but they let it slide.

Months later, the soap company returned. They now had a specially formulated hand soap for extra dirty jobs. Basically, shop soap. They wanted to do the same thing: Contribute advertorial blog posts written by them. This time, I was more intrigued with the product and thought we could come up with an advertorial campaign that made everyone happy. We had wanted to start doing more maker profiles on the site. So, I proposed that we do a series of profiles on makers whose work was particularly filthy (blacksmithing, an artistic residence at a city dump, car customizing, smelting, etc.). We would write the pieces ourselves and, at the bottom of each profile, there would be a set-off section that said the content was sponsored. And there would be a logo and a link to their dirty jobs soap campaign website. The sales team and the company loved it. This was exactly the type of content that we would have normally run and the advertisement at the bottom was non-invasive. I always counted the success or failure of one of these sponsorship campaigns by how much negative feedback we got in the comments. This campaign received zero. Everyone was happy. And no content was harmed in the process of making a buck.

⭐ KNOW YOUR TARGET!

So many YouTube makers are now using selfie sticks and GoPros while doing projects on a bench, with their camera aimed at the work. Cheap, hi-def cameras have been a boon to content creators. But having a camera above your work or holding it as you move around can be hugely frustrating to the viewer if you don't stay within frame. I look at a lot of YouTubers' videos and am shocked by how often the workpiece or other target for the shot goes off-frame. Make sure that you know the boundaries of your work area before you start filming and stay within that target. Always.

⭐ PLEASE, PUT YOUR NAME ON YOUR WORK!

This has been a pet peeve of mine for years, and I have written about it numerous times. Unless you're in the Witness Protection Program and don't want people knowing who you are, please put your name in your YouTube bio and/or in video descriptions. People want a name to put to the face. It personalizes and grounds your content. You're trying to build channel/brand loyalty, and you are the brand. And for those of us who write about and try to spread the good news about your great content, we don't want to have to spend 20 minutes trying to find your name so that we can refer to you in an article or blog post.

⭐ CLEAN YOUR FINGERNAILS (WHEN POSSIBLE)

When someone's fingernails are filthy and their hands are dirty, it can be off-putting to some viewers. Now, if you're doing blacksmithing or auto mechanics or letterpress printing, it would be weird if your hands weren't dirty. But if you're sewing or bookbinding or soldering or some other relatively clean work activity, it's nice to have your hands clean and your nails trimmed. Oh, and the same goes for dandruff on your shoulders and the day's lunch in your teeth. Check yourself in a mirror before you roll the camera.

MUST-SEE MAKER TV

If you watch any DIY content on YouTube, you're likely already well aware of folks like Jimmy DiResta, Adam Savage, Laura Kampf, Donald Bell, Simone Giertz, Izzy Swan, and April Wilkerson. Here are some of my favorite channels that are a little more "deep cuts" and may not already be on your radar. This is by no means an exhaustive list, just a few of the channels I think readers of this book may want to check out. You can look up any of these makers by searching for their channel name on YouTube.

THE 8-BIT GUY

I've been a fan of *The 8-Bit Guy* (David Murray) for years. I'm a sucker for retro-computing and antique tool restoration, and these come together in *The 8-Bit Guy* projects. Watch and learn as David brings old PCs, Macs, Commodore 64s, Amigas, game systems, and electronic toys back from the dead.

ALEX

French chef Alex Gabriel Ainouz, of the YouTube channel *Alex*, combines one of the most basic forms of making (preparing food) with his training as an engineer to create a fascinating take on the art and science of cooking. He can't just make pasta noodles; he must build his own pasta machine first. His quest for the perfect meatball takes him on a trip around the world. His food and kitchen tool explorations are always as entertaining as they are educational.

BILL MAKING STUFF

If deadpan comedian Zach Galifianakis (*Between Two Ferns*) had a hobby crafting channel, it would look something like Bill Mullaney's *Bill Making Stuff*. In a similar flat, flippant, and funny style, Bill shares his love of art journaling and building fantasy and sci-fi models from scratch and from junk. While his channel is geared towards modeling for tabletop gaming and display, many of his tips and techniques can be applied to many forms of crafting.

BLONDIHACKS

Quinn Dunki is a game designer and machinist who covers electronics, homebrew computing, metalworking, and machining on her channel. Her milling and machining videos are really valuable and informative to newcomers and seasoned metalworkers alike.

CHRIS NOTAP

I love YouTube channels that are "little more" than clever DIYers letting us follow along as they explore, repair, hack, and improve the world around them. One of my favorite examples of this is Chris Notap. He perfectly summarizes his channel as "Inventive - Random - Creative." Chris covers everything from home improvements and repair to cooking and kitchen hacks to literally building a better mousetrap. And, it's all delivered in a very clear, thoughtful, and patient production. Good stuff.

CLICKSPRING

There is a breed of maker whose level of skill is so high, their channel so expertly produced, that watching them is awe-inspiring (if not a little bit intimidating). Chris Budiselic of *Clickspring* is such a virtuosic maker. His channel focuses on clock-making technology, and he fashions many of his tools and watchworks entirely by hand. Chris is best known for a multi-year project to recreate the Antikythera mechanism, a Greek bronze age astronomical computer, parts of which were discovered amid an Aegean Sea shipwreck in 1901. He is reverse engineering the device using as much technology as possible that would have been available to its ancient makers.

CNC KITCHEN

On *CNC Kitchen*, German 3D printing enthusiast Stefan Hermann explores the science and engineering behind 3D printing. He looks at things like filament strength, component wear, how object designs impact print quality and strength, the strength of different infill designs, and more. He also reviews printers.

DINAA AMIN

Dinaa Amin is an Egyptian maker and product designer who, among other things, shoots amazing stop-motion videos of product teardowns.

GREATSCOTT!

I have long been a fan of the YouTube channel *GreatScott!* This young German maker produces really excellent, deep, detailed, but surprisingly clear projects and tutorials on electrical engineering and working with circuits, ICs, and microcontrollers.

HAND TOOL RESCUE

Eric of *Hand Tool Rescue* makes magic happen on his channel as he turns filthy, rusted, antique tools and machines into something as close to showroom-new as he can possibly achieve. And he does it with moments of laugh-out-loud foolishness and fun.

LUKE TOWAN

I cannot get enough of Luke Towan's eponymous channel. His skills as a modeler are unmatched, and he has such a low-key, confident manner that he makes hyper-realistic train board modeling look easy and effortless (it isn't). The techniques he uses apply far beyond train layouts. Tabletop game modelers, dungeon crafters, diorama artists, movie F/X modelers, and others will find tons of inspiration here. Even if you have nothing to do with any of these crafts, you may find these videos as compelling as I do.

MR. CHICKADEE

I just recently discovered the wonderful *Mr. Chickadee*. On it, "Mr. Chickadee" lets us follow along as he and his wife create a beautiful rough-hewn homestead in the wilds of Kentucky using little more than antique hand tools and traditional methods of carpentry. The unnarrated videos are beautifully shot by Mrs. Chickadee. Mr. Chickadee is one hell of an impressive old-school carpenter.

PNEUMATIC ADDICT

Elisha Albretsen, aka *Pneumatic Addict*, is a maker who specializes in simple, modern home decor and custom furniture builds. She and her husband just built their own impressive modern-style home in Arizona.

THE POST APOCALYPTIC INVENTOR

Is it just me, or do preppers and doomsdayers seem more reasonable in their obsessions these days? I'm certainly starting to pay more attention to fundamental survival, maintenance and repair, as well as traditional shop craft content. Ya know, just in case. One YouTube channel I've recently been binging on is *The Post Apocalyptic Inventor*. German maker Gerolf Kebernik scrounges through high tech (and low tech) junk and does wonders with diagnosing, fixing, and bringing machinery back to life. I've learned (and already applied) countless machine repair tips in the short time I've been watching.

PRACTICAL ENGINEERING

Watching educational programming as good as Grady Hillhouse's *Practical Engineering*, it's easy to underestimate just how hard it is to create something this watchable on subjects as technical and complex as the ones he so deftly tackles. In each (usually) 8-to-10-minute episode, Grady looks into some aspect of the physical world around us, from water towers and sinkholes, to dams and the wonders of Roman concrete. Besides the thumbnail education on each topic, Grady also performs clever and eye-opening lab experiments to clearly illustrate the principles being discussed. Really impressive stuff.

SEE JANE DRILL

Leah Bolden is a building contractor and educator. On her channel, *See Jane Drill*, she goes over tools and techniques of the trade that are relevant to both home DIYers and to anyone who wants to know more about things like tape measures, rulers, carpenter's pencils, caulking, and more.

SOPHY WONG

Sophy Wong is a designer and maker who explores all sorts of curiosities and passions on her YouTube channel. She works in electronics, high-tech wearables, cosplay, VR/AR, 3D printing, and more. Sophy is also sometimes a guest host of Donald Bell's weekly must-see YouTube show "Maker Update."

XIAOQIANFENG

Xiao Qianfeng is a Chinese maker who lives in the Netherlands. She creates the most beautiful costumes and everyday items (suitcases, shoulder bags, chairs) mainly out of paper, cardboard, and other reclaimed materials. Her videos are beautifully shot and paced, lyrical, and the resulting objects are often jaw-dropping.

XYLA FOXLIN

An engineer with a passion for art and adventure, Xyla Foxlin, describes herself and her channel as "a bit of a speedy jet with a broken compass. I'm always racing in the direction of the skill that's caught my interest... for now." Xyla explores everything from aviation to wearable art to science experiments with exuberance and curiosity, with impressive results.

KITCHEN TIPS

Makers have got to eat. And (at least sometimes) cook what they eat. Creating in the kitchen is one form of making that most of us share. Like other forms of making, it requires some level of planning, tools, materials, supplies, and tried and true techniques. And all that is made better, more efficient, creative, and fun with useful tips and tricks of the trade. What follows is a collection of tips that should serve anyone who cooks and likes to create in the kitchen.

★ BURGER STACK HACK

How often do you get a restaurant burger, or grill one yourself, and before you're finished horking it down, the soggy bun has lost the will to live and disintegrates in your hands? Here's my fix for your fixins. Don't let the meat or condiments touch the bun directly. Instead, create a protective shield with the lettuce, and stack the meat and condiments underneath. No more soggy burgers.

★ SHAVING OFF BURNT COOKIE BOTTOMS

If you burn the bottoms of cookies a little, you can shave the burnt material off by using a rasp-style grater. In a pinch, you can just scrape the bottom gently with a regular ol' butter knife. Our oven runs a little hot, and, if you take your eyeballs off the baked goods for a second, they can overcook. This trick has saved us on more than one occasion.

⭐ HAMMERS IN THE KITCHEN

My pal and well-known YouTube food geek Alex Gabriel Ainouz, of *Alex*, recently teamed up with blacksmith Alec Steele to create his own meat hammer. Once he got it back to his kitchen in France, he decided to experiment with all of the various ways one can use a hammer in the kitchen. Besides obvious things, like tenderizing meats and flattening them,he also used it to hammer a knife through bone and frozen foods, to crush ice, garlic, and nuts, and to flatten plantains for frying. The moral of the video? Get a hammer/mallet for your kitchen and see what uses present themselves. **[AGA]**

⭐ PITTING OLIVES WITH A FUNNEL

An easy way to pit olives is to turn a funnel upside down so that the wide mouth of it is on the counter. Press your olives over the throat of the funnel and the pits will fall inside. Neat!

⭐ BREAKING DOWN HARDENED BROWN SUGAR

We all know this deal. You buy a box of brown sugar for a recipe. You go to use it a few months later and the sugar has become a sweet, brown brick. One great way to turn that sugar-brick back into usable granules is to grate it on the large-grate side of a box grater or a flat hand grater.

Juliann Brown

⭐ BINDER CLIP SPONGE

Sponges can quickly get sour and funky sitting wet on the sink. You can fix that and give your sponges a longer life by attaching a binder clip to one end. Splay out the two handles on the clip to act as feet and stand the sponge up after squeezing it out. This allows it to dry without one side remaining damp against the sink or counter surface. You can even do most tasks with the sponge without having to remove the clip.

⭐ GETTING CORK "CRUMBS" OUT OF WINE

It's inevitable that you're going to have floaters once in a while, pieces of cork that end up in your fancy (or two-buck Chuck) bottles of wine. An easy way to get them out is to use a straw. Stick the straw down into the bottle's neck and place the end of it over a cork crumb. Place your thumb over the other end of the straw to create suction. Remove the straw and lift your thumb to release the crumb. Repeat to remove any additional crumbs. Pour and enjoy.

⭐ PUT YOUR UNUSED TRASH BAGS IN THE TRASH

When I talk about tips, I'm always interested in the ones that stick. I read about this tip in *Family Handyman* magazine and have been using it ever since. I have a "thing" about emptying the trash (perhaps childhood trauma over being the family garbage man). I hate it. For a while, I was triple-bagging the trash can (so that you

peel off the inside, full bag and you have another all ready to go).
But then I read about just leaving the whole roll of plastic bag liners
in the bottom of the can so that you can slide a new one on after
you remove the full one. Game-changer.

⭐ DIP, DON'T POUR, YOUR SYRUP

I discovered this little trick myself. I wondered if dipping your
pancakes or waffles into the syrup, rather than pouring the syrup
over them, would save a significant amount of syrup. It does! I did a
series of (delicious) experiments and discovered that you save
about half as much syrup by dipping instead of pouring. And, for
those of us who insist on genuine tree blood (er... maple syrup), this
can save a lot of money (and you're not eating more sugar than you
need to).

Adobe Stock - ahirao

⭐ ORDERING DELIVERY FOOD LATE?

I'm not sure if this is a real tip or not. Sometimes, working late in my
shop, I have ordered delivery food right before the restaurants
close. Each time, I've noticed that the portions seem to be greater
than normal. In one Chinese order, the delivery bag weighed a ton
and the boxes were crammed with food. I assume that, within an
hour of closing, the restaurants are trying to get rid of any
remaining food that they're likely just going to throw out. I haven't
done this enough to see if it's a reliable pattern. Feel free to
experiment and report back.

MISCELLANEOUS

The word "miscellaneous" (which dates from around 1630-1640) is from the Latin *miscellāneus*, meaning "mixed, of all sorts." For *Tips and Tales from the Workshop*, Volume 2 (2022), it means "a mixture of great tips that didn't fit into the other chapters of this book." See also: "grab bag."

⭐ CUTTING DOWN ON AMAZON WASTE

If you're as disgusted as I am by the phenomenal waste of Amazon packaging, there are a few things you can do to minimize it. Besides the obvious recycling of shipping cartons, you can flatten the air pillows and recycle them in the plastic bag recycling bin at your grocery store.

Did you also know that Amazon will provide free shipping labels (givebackbox.com) for you to use one of their boxes to fill up with charitable items that you want to donate? Another thing you can do to cut back on package waste is to wait. Don't order things piecemeal. If you can wait, place things in your cart and order at the same time. This increases the chances that the items will be shipped in one or at least fewer boxes. Amazon now offers a Monday delivery service where you can set things aside as you order them to all arrive in a Monday shipment.

To find out more about Amazon recycling, product trade-in, and buying refurbished and returned products, check out the Amazon Second Chance program (amazon.com/amsc).

⭐ KNOWING THE LIMITS OF YOUR BODY WHEN DRAWING

I've recently gone down the rabbit hole for an indie tabletop miniatures game called "Relicblade." Besides being an amazingly cool fantasy skirmish wargame, "Relicblade" is also special because it is primarily the vision and labor of one maker, artist, and game designer: Sean Sutter. Sean pretty much does everything, from game design to art and illustration, design and digital sculpting of the miniatures, book design — all of it. And he documents much of his process on YouTube. In one of those videos (youtu. be/1eQZ5PCetmE), on conceptualizing and rendering his thief character for the game, he shares a very important concept that can apply to all sorts of making, but especially anything where hand and arm control are important (as in drawing). Sean says to think of your hand/arm as a machine, with mechanical capacities and limitations. Once you are mindful of these, you can work around/with them. In doing an activity, in Sean's case drawing, you want to think about the optimal angles for confident control of your hand and to move the workpiece to conform to your limitations rather than trying to make your hand do things it doesn't naturally want to do. You can see in the above-linked video that, in drawing his thief image, Sean is constantly turning the paper to get the best, easiest angle. [SS3]

⭐ WHEELBARROW RECLINER

That moment when you realize that your wheelbarrow can also be used as a surprisingly comfy chair (cold beer optional). From the Homestead/ Survivalism Facebook group.

⭐ HOW TO TIE YOUR SHOES

YouTuber Dirt Farmer Jay offers this tip on tying your shoes to help prevent the bow knot from "spilling." The idea is simple: Place the left lace over right lace instead of right over left (as is most common). You can also double-loop the bow knot for extra security. **[DFJ]**

⭐ COFFEE MUG BAROMETER

A reader of the first volume of *Tips and Tales from the Workshop*, Joe Mayerik, was inspired to send me some tips. Here's one: Instructables user stickmop shares a trick to tell basic barometric pressure using a hot cup of coffee, tea, or cocoa. When you pour the coffee into the cup, watch the bubbles. If they move to the edge of the cup quickly, you can expect clear skies for the next 12 hours. If they hang out in the center, expect rain in the next 12 hours. And if the bubbles move slowly to the edge, you may get a bit of weather, but it will clear soon. This trick came from a long-ago issue of

Backpacker Magazine. If you're out camping without an internet connection, it may come in handy. **[JM2]**

⭐ CHEAP AND EASY ELECTROPLATING

On his YouTube channel, Geoffrey Croker did a piece on how easy it is to do basic electroplating. All you need is a container, like a glass canning jar, a power source (two D-cell batteries will work), wires with alligator clips, some copper wire, white vinegar, salt, and a piece of metal you want to plate. He recommends starting with nickel. You can find strips of nickel (and other metals) on eBay by searching on "nickel anode." Or you can find other suitable metals by searching for "copper" or "zinc."

Fill the container with vinegar and add a tablespoon of salt. Cut your strip of nickel in half lengthwise and bend each piece over the container. Connect the power supply to each piece of nickel. After a few minutes, look at the two nickel strips. Note which one has a lot of bubbles forming. That is the negative side. In about two hours, the solution will turn green. Congratulations, you have successfully created a nickel electrolyte. Other metals will turn the solution different colors. Copper electrolyte will be blue. Zinc, a cloudy clear.

Now, to prepare your metal pieces for plating, you first need to get them immaculately clean. Suspend the part to be etched from a copper wire. If you have some hydrochloric acid (mixed 50/50 with water), wash the pieces off first before suspending them in the electrolyte. This is optional. If you do use it, wash the pieces to be plated off in water before suspending to remove the acid.

It's now time to do some nickel plating. Put a nickel strip over the edge of the electrolyte and attach the negative power wire to it. Now, suspend the piece to be plated in the solution using a dowel or carpenter's pencil or similar. Wrap the copper wire around your pencil (as shown in this illustration) and leave the end of the wire sticking out. Attach the positive power wire clip to it.

In about 45 minutes, you'll have a nickel-plated piece. Copper and zinc plating work the same way. For copper, first nickel-plate your part in nickel electrolyte and then plate in a copper electrolyte bath that you have prepared in exactly the same way as you did the nickel bath. Once you've created these electrolytes, you can keep them and use them over and over again as long as they're in an airtight container. See Geoffrey's video for more information: youtu.be/G-PtnwtOR24. **[GC]**

⭐ ETCHING WITH A 9V BATTERY

On her YouTube channel, Leah Bolden of *See Jane Drill* shows how easy it is to etch a metal surface using a 9v battery and wires with alligator clips, vinegar, salt, and Q-tips.

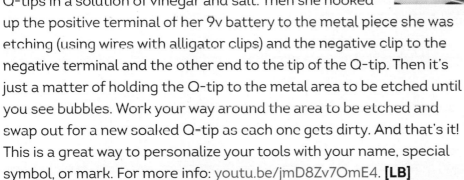

For her resist (the area that she doesn't want etched), she used some vinyl letter carrier material to spell out her name. To do the etching, she first soaked some Q-tips in a solution of vinegar and salt. Then she hooked up the positive terminal of her 9v battery to the metal piece she was etching (using wires with alligator clips) and the negative clip to the negative terminal and the other end to the tip of the Q-tip. Then it's just a matter of holding the Q-tip to the metal area to be etched until you see bubbles. Work your way around the area to be etched and swap out for a new soaked Q-tip as each one gets dirty. And that's it! This is a great way to personalize your tools with your name, special symbol, or mark. For more info: youtu.be/jmD8Zv7OmE4. **[LB]**

⭐ FIRE-STARTERS FROM TOILET TUBES AND DRYER LINT

I don't know about your dryer, but mine generates a lot of dryer lint (as my clothes get smaller and smaller and smaller). Here's a not-so-silly idea for using this lint and toilet paper tubes to create fire starters. Just stuff the TP tubes with lint and keep them by your kindling and firewood.

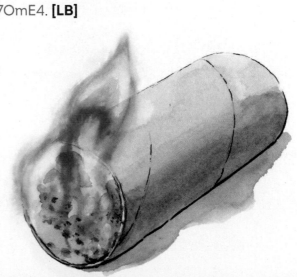

⭐ CHANGE FONT WHEN PROOFREADING

I love this tip from Derek Thompson (@DKThomp), a writer for *The Atlantic*, that he shared on Twitter: "Simple copy editing tip from somebody who sucks at copy editing: In your last pass, change the font to something unfamiliar. Then change the font size. When you're familiar with a piece, your eyes gulp whole passages and you can miss typos. New fonts refocus your eyes on each letter." **[DT]**

⭐ PRINT TO EDIT YOUR WRITING

When you're ready to do a deep edit on something you've written, print it out, move to another space, and read the piece that way. Reading it in print really changes your perspective on it. Bonus: Read it out loud. It will help you to compose a smoother flow.

⭐ PROTECTING HAND-ADDRESSED LETTERS

For years now, especially during winters and rainy seasons, I always make sure to cover my hand-addressed letters and package labels with clear shipping tape. That way, if the envelope gets wet, the ink won't run and obscure the address.

⭐ DESIGNING A LOGO

During the Q&A of my World Maker Faire talk with Jimmy DiResta, someone asked us about considerations for designing a logo for a YouTube channel/maker biz. Since my first career was as a graphic designer, I had plenty to say on the subject. I offered several suggestions. When you get some logo designs that you like, try enlarging them huge and then reducing them tiny. Make your logo bold and emblematic enough that it will look good at any size. Avoid the temptation to say too much (literally or figuratively) on your logo. Get a flyer for an event that has a bunch of the sponsors' logos on it (usually printed very tiny), reduce your proposed logo to that size and see how it looks amongst the others. Also: Avoid trendy treatments and typefaces that will date your logo going forward.

⭐ STENCIL-MAKING APP

Via Donald Bell's "Maker Update" YouTube show comes news of a very useful little graphics app called Stencilfy (stencilfy.herokuapp. com). All you do is type in any text, and the app will render it in stencil form. It has three default fonts and you can upload fonts from your computer. You can also save your stencils as Scalable Vector Graphics (SVG) files. **[DB]**

⭐ BALL VALVE ON OR OFF?

On an Instagram story, Jimmy DiResta offered this handy little reminder (while haunting the aisles of Home Depot): "For any ball valve that has a cut off, when the lever is in-line with the pipe, the valve is *On*. When the lever is perpendicular to the pipe, it is *Off*." **[JD]**

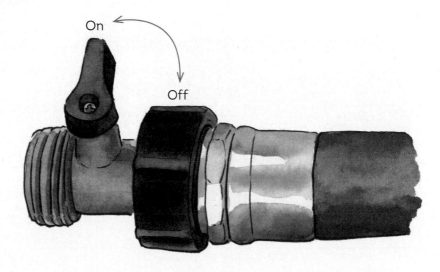

⭐ POOL NOODLE BIKE SPACER

Kevin Kelly tweeted about this clever idea for cyclists: strap a foam pool noodle on the back of your bike to create a bright visual alert for motorists and to indicate a safe passing distance. **[KK]**

⭐ DECOMMISSIONING A HARD DRIVE

As Dan Hienzsch reminded us in a tweet, there is only one way to effectively secure the data on an old hard drive that you're throwing out, and it involves a power drill and a serious bit. Find the area inside the computer case that you are throwing out where the hard drive is located, or a flat side of a removed hard drive, and sink a few holes into it with your drill. Otherwise, your info could easily end up in the hands of some enterprising data thieves halfway around the world. **[DH]**

⭐ ZIP-TIE CABLE TIES

Newsletter reader Keith Monaghan writes: "I used zip-ties and #6×½" screws to make little harnesses for the drip lines that water my wife's hanging baskets. I wrapped the ties around the drip line leaving them one click looser than snug. I then positioned the zip-ties and drove a screw through the long 'tail' of the tie, then trimmed it off. It works for runs of wires in the garage, too. I'd use wider zip ties or smaller screws next time, as the ones I used were a bit tricky to properly align. It's a quick and dirty solution that will last through summer and until I can come up with a more discreet path for the drip lines next spring." **[KM]**

MORE TIPS AND TALES: LET'S TALK

Ultimately, this books is really about community, and conversation, and the stories that are attached to the tools and work techniques that we use to build our lives. So, let's talk. If you have a tip, a favored tool or technique for any manner of making, please tell me about it. If you have stories that go along with your tips and tools, I'd love to hear those, too. Your tips and tales may end up in a future tips column of mine or in a future edition of this book. You can email me at garethbranwyn@mac.com. And, please sign up for my free newsletter to get a weekly dose of tips, tools, and shop tales: garstipsandtools.com.

INDEX